DESTINATION
INNOVATION

HR's Role in Charting the Course

W0038452

DESTINATION INNOVATION

HR's Role in Charting the Course

Dr. Patricia M. Buhler

Society for Human Resource Management
Alexandria, VA
shrm.org

Society for Human Resource Management, India Office
Mumbai, India
shrmindia.org

Society for Human Resource Management
Haidian District Beijing, China
shrm.org/cn

Society for Human Resource Management, Middle East
and Africa Office
Dubai, UAE
shrm.org/pages/mena.aspx

Founded in 1948, the Society for Human Resource Management (SHRM) is the world's largest HR membership organization devoted to human resource management. Representing more than 275,000 members in over 160 countries, the Society is the leading provider of resources to serve the needs of HR professionals and advance the professional practice of human resource management. SHRM has more than 575 affiliated chapters within the United States and subsidiary offices in China, India and United Arab Emirates. Visit us at shrm.org.

Interior and Cover Design: Mari Adams

Library of Congress Cataloging-in-Publication Data
Buhler, Patricia M.
Destination innovation : HR's role in charting the course / by Patricia M. Buhler.
    pages cm
Includes bibliographical references and index.
ISBN 978-1-58644-383-2
1. Personnel management. 2. Technological innovations—Management. 3. Creative ability in business. I. Title.
 HF5549.B87357 2015
 658.3'01--dc23
                    2015008578

ISBN: 978-1-586-44383-2   15-0131

Contents

Dedication..viii

Chapter 1. An Introduction to Innovation: More Than Thinking

    Outside the Box 1

Chapter 2. Understanding the Foundation of an Innovative Culture:

    Making It Work!.. 13

Chapter 3. Hiring for Innovation: Feeding Our Culture.......................... 33

Chapter 4. The Awakening: Developing a Workforce that Embraces

    Innovation 49

Chapter 5. The Carrot and the Stick: Rewarding for Innovation.......... 61

Chapter 6. Managing People in an Innovative Culture........................... 73

Chapter 7. The Shoemaker's Children: Is HR's House Innovating? 89

Chapter 8. Parting Thoughts: Putting It All Together 101

Endnotes 117

Index 131

About the Author...137

Additional SHRM-Published Books...139

This book is dedicated to my family. I am so grateful for the enduring support and unconditional love of my husband, Jim, and my children, Jennifer and Matthew. Thank you for always believing in me.

Chapter 1.

An Introduction to Innovation: More Than Thinking Outside the Box

Our ability to innovate affects our organizations' performance—regardless of company or industry. One survey found that 98 percent of its respondents reported innovation as either important or very important in the success of their firm going forward. Furthermore, those companies that possess an openness to change and that focus on innovation tend to be the market leaders.

The business case for innovation and an expanded understanding of what innovation means today are explored in this chapter with surveys and research reports providing the foundation for the growing importance of innovation today—across all industries and organizations. Yet perhaps the strongest case for the importance of innovation is found in the best practices (and business performance) of successful organizations.

The nearly impossible can become possible with innovation, especially when executed as a team. The rescue of the 33 trapped Chilean miners in 2010 is a widely publicized example of the power of innovation. Harvard Business School professor Amy C. Edmondson shared that "it took the collaborative efforts of over one hundred experts in diverse fields innovating to develop and execute a novel solution on the fly."[1] Consider the possibilities in our own organizations with a commitment to innovation!

Why Innovation?

To generate a competitive advantage in today's business landscape, companies must continually change. In their book *Competing for the Future*, Gary Hamel and C. K. Prahalad sent the clear message that organizations must be fast, agile, and responsive.[2] We cannot be slow to take action. The window of opportunity is open for shorter periods of time today as our time horizons diminish. And being agile means being flexible, not being wed to the past.

Spotlight on Innovation

"As an organizational capability, agility represents a virtuous cycle. It enables an organization and its workforce to be more innovative, to think critically, and to continuously learn and improve."[3]

The "same old, same old" (or business as usual) simply does not cut it today. In fact, there is no "usual" in today's dynamic business landscape. Those organizations that are not moving ahead are quickly falling behind. And in a rapidly changing world, falling behind just a little often puts us in the precarious position of playing the eternal catch-up game—until alas, we can no longer catch up, and we perish.

The firm that embraces innovation is better positioned to be responsive to its customers, anticipating their needs and responding accordingly. The innovative organization can often identify what their customers want before the customers even know it. How many of us cannot imagine a day without our smartphones? Apple may have very well envisioned this day far before we did!

Matt Donovan, author of the article "Shifting Focus to Agile Development" made the point that agility requires more systematic change rather than simply addressing pockets of change.[4] An agile manifesto that was written at a computer programmers' summit in 2001 can provide a refresher for us today in moving away from the more traditional approaches to doing things. Much of the change so necessary for today's world, then, may require firms to reinvent themselves. And at the heart of this ability to change is the very DNA of the firm, known as its organizational culture. A culture that embraces innovation is critical in addressing the need for change.

According to Michael Stanleigh, "A recent study by the Harris Group indicated that executives see a culture of innovation as crucial to not only growing their business (95 percent) and profitability (94 percent) but also for attracting and keeping talent (86 percent)."[5] It is no wonder, then, that the new battle cry of business has become "innovate or die." Recent surveys have reported that organizations across the globe are placing increasing importance on the need for innovation—and rightly so. The ASTD white paper, *Building an Innovative Organization: The Role of Training and Development*, reported,

> "Today's executives firmly believe that innovation is central to a company's strategy and performance, but getting it right is as hard as ever. . . . PricewaterhouseCoopers (PwC) even stated that 74 percent of CEOs regard innovation as at least equally important to operational effectiveness."[6]

Spotlight on Innovation

"Innovation is a collaborative process; where people in many fields contribute to the implementation of new ideas."[7]

Depending on the criteria and which publication or company is generating the list, we see a variety of companies across a wide range of industries named as 2014's most innovative companies. *Forbes* highlighted Salesforce.com, Amazon.com, Regeneron Pharmaceuticals, and Unilever Indonesia among its top picks. We see recognizable names such as Coca-Cola and Chipotle Mexican Grill on others. Boston Consulting Group (BCG) has named Apple as its top pick for nearly a decade. Google, Microsoft, and IBM remain among the top 10 year after year.[8] Samsung has been climbing in the rankings over the past few years. And two automakers made the top 10 this year. We see that it is not just about tech companies. The question we need to ask is, "What is it that they know and do that we don't?" Perhaps first and foremost, they recognize the importance of innovation. Do you? Do your organization's leaders even understand what it means to be innovative?

What Does It Mean to Think Outside the Box?

Creative thinking is an essential element of an innovative culture. Everyone must embrace creativity to think outside the box. But what does this really mean? The term "thinking outside the box" comes from the classic nine-dot exercise where we are asked to connect all nine dots in a 3 x 3 configuration with four lines while never lifting our pencils from the paper (see Figure 1.1). You might try this before peeking at the solution at the end of the chapter.

Figure 1.1. The 9 Dot Challenge

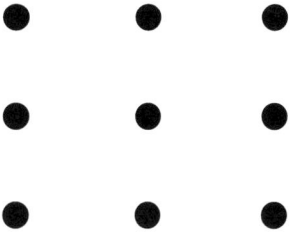

One of the greatest barriers to thinking creatively is demonstrated in this challenge, that is, placing artificial constraints on ourselves—and on our thinking. When we are hindered by these constraints, we cannot solve the challenge of the nine dots because our attempts are always made when drawing the lines inside the box created by the nine dots. When viewing the solution (see Figure 1.2 at the end of the chapter), we can see that it requires drawing lines beyond the limits of the nine dots. And thereby we learn to literally "think outside the box."

Too often when faced with a challenge, we place these artificial constraints on ourselves and fail to truly be innovative. Instead, with our limiting assumptions we simply tweak our existing approach to the challenge with only an incremental improvement.

A New Perspective on Innovation: What and Where

But what exactly does it mean to be innovative? For decades we talked of building the proverbial better mousetrap. Creating a sticky surface on the traditional trap produced a better mousetrap. However, "better" is not enough for today's innovation. This is simply an incremental improvement. But if we designed an entirely new approach to ridding our houses of mice that did not involve a trap at all, *that* could be innovative. Tinkering and tweaking are not the real innovation that our organizations need.

Today's innovation requires that we challenge our assumptions and test what we think we know. Although traditionally many of us may have thought of innovation as "charting new territory" so to speak, the current perspective also involves putting knowledge together in totally different ways.

Innovation is *not* just copying what someone else is doing, but it may be "copying" or using it in a new way or in a new department. There is value in looking outside our industries given that most of the technology that disrupts industries comes from outside our own.

In his book *The 3M Way to Innovation*, Ernest Gundling shared 3M's definition of innovation as "more than just a bright idea; it is an idea that gets implemented and has real impact. In other words, somebody has to make it happen."[9]

Until recently, we have focused primarily on product and process innovation in the manufacturing environment. That is, we focused on the design of new products to offer our customers, or we focused on the design of new processes to create those products for our customers. Today's definition of innovation expands beyond these.

We see common agreement in the new definitions of innovation that embrace this broader view beyond new product and process development. Claude Legrand, managing partner of Ideaction, broadly defined innovation as "the process of identifying new solutions that create new value for some or all of the stakeholders of an issue."[10] Author and professor David Burkus suggested that "innovation means more than just new products or services. It means improving the process of creating those products, or selling them, or experiencing them, or even improving the ways we manage the people who do all of the above."[11] Products, processes, and people are all included in this current, expanded view of innovation.

We can agree that this innovation is not relegated to our engineers or confined to our operations and marketing professionals. This innovation, then, is organizationwide. It includes everything and is everywhere! And most importantly, it involves everyone! For example, "BMW ensures that all departments are focused on innovation. While some organizations focus only on manufacturing, BMW also focuses innovation on every department within the organization including sales and marketing, human resources and product development."[12] To be effective, we must agree that innovation is everyone's business.

Spotlight on Innovation

"Tomorrow's business heroes may not do things in an expected way, and may look different than you expect—your company may not be ready for pink hair, nose studs and leather boots, but allowing deviations from the norm when you can is a step in the right direction. You want an organization that does things differently, that thinks differently? Then be different."[13]

Today's companies are most likely to make the initial move into the innovative mode by focusing on changes in processes and technology. There is a missed opportunity, however, when the human element is forgotten. Building in time to think and incorporating opportunities for collaboration help address the human factor in innovation. And let us not forget the management processes needed to support the culture. We must think innovatively here as well.

Sustaining the Engine of Innovation

We can effectively address the challenges faced by our organizations, our industries, and our countries only with innovation. This is certainly no management fad and no "flavor of the month." Innovation is the key to the solutions for organizational survival and growth that we so sorely seek. It can very well be the best approach to the achievement of our goals and may just be the source of our competitive advantage.

A great example of an organization whose competitive advantage is based on its commitment to innovation is provided by Apple. While other companies were answering the question of whether to choose a focus on quality or a focus on cost, Apple said yes to both. And then it delivered on its promise. Yet the delivery on this promise has not happened by accident. The commitment to innovation is companywide. It does not belong to one department. As much thought and design went into the packaging for the iPhone as for the phone itself. And even the work environment itself represents a deviation from the traditional corporate environment. Innovation belongs to all employees and permeates all they do. It is not just a goal for Apple's products; innovation is embraced in its business models as well. This comes from a companywide commitment to the value of innovation. This commitment is woven into the very DNA of the company—into its culture.

Harvard Business School professor Rosabeth Moss Kanter suggested that "innovation gets rediscovered as a growth enabler every half-dozen years. . . . However, grand declarations about innovation are followed by mediocre execution that produces anemic results."[14] This is no longer a fad that we can address every few years. This is an ongoing commitment. We must constantly challenge our assumptions, question, and ask "why not?" Innovation requires that we test what we think we know.

As a continuous initiative, innovation must be embraced in our culture and reflected in our organizational values.

Kanter identified multiple waves of innovation beginning in the late 1970s that were essentially responses to the current business challenges and changing landscapes. The common theme among these waves has been the tendency to repeat the same mistakes. This is where we must break from the past—we must execute beautifully and continuously.

Kanter noted that an inability to be courageous is a critical mistake. Seeking innovation and then looking around at other companies to see if others are doing the same thing we are is self-defeating, and certainly not courageous.

Spotlight on Innovation

"The ability to help create, protect and build organizational culture is a critical role for HR to play, as it is a major driver for innovation."[15]

In good economic times, fewer companies engage in innovations that are more disruptive in nature. Instead, we are lulled into a sense of complacency or inertia due to our "good times." However, if our businesses are threatened or if economic times are no longer good, *then* we want to invest in new growth opportunities—we want to innovate. The key, though, is to launch new initiatives while we are still growing and doing well. When we are desperate, we are not patient to grow ideas, and our implementation is more likely to be flawed.

Innovation must be woven into the very fabric of the organization. It is not just about the products and services offered. Alex Gammelgard, director of product marketing at Apttus, suggested that to foster an innovative culture, we do not have to recruit geniuses.[16] We do, however, have to provide the opportunities for our employees to build their self-confidence and reward their efforts in finding solutions to problems until innovation is deeply ingrained in the culture of the organization.

Understanding the Path to the Innovative Organization: When, How, and Who

Let's be clear. Innovation is everyone's job. To be effective as an innovative organization, all employees must accept the responsibility to innovate—to see everything they do and everything around them through a new lens.

Innovation requires a commitment to reinventing our companies. This extends beyond just products. Instead, it encompasses *all* aspects of the business. Perhaps the biggest caution for organizations today is that we cannot wait until something is broken. Reinvention must be started early. Paul Nunes and Tim Breene remarked that we must "learn to focus on fixing what doesn't yet appear to be broken."[17] Though this seems like simple advice, it is difficult at best to implement.

In addition, we must rethink the scope of our innovation. We have to move away from funding only the big impact "home run" opportunities and instead take a chance on the ideas that still need some shaping. This is usually in the realm of middle managers, who tend to be cautious (often because of the misunderstood risk to their careers).

We do not want to go just for the big ideas with the big payoffs. We make a mistake in overlooking opportunities that at first glance seem small, that is, not big enough. Effective innovative practices recognize that the small wins are important. There is a compounding effect of combining many "small win" innovative approaches. Innovation of all sizes should be explored!

Implementation: Connecting the Dots

Companies seem to understand the importance of innovation, but they continue to experience confusion on the execution side. That is, there is an acceptance of the need to innovate and an eagerness to read about best practices, but it is still not clear to many how to connect the dots—to understand how innovation directly contributes to the bottom line of the organization and how to integrate it throughout.

Some business people have suggested that innovation is as elusive as the holy grail, and they have even drawn a parallel to its importance. So although we understand the need for innovation, we need to understand how we can foster that innovative culture—that organizational environment that enables all employees to accept that innovation is their job and then execute on that belief.

Spotlight on Innovation

"An IBM Global CEO study in 2008 cited an unsupportive culture as the number one obstacle to innovation."[18]

For decades, 3M has been held up as the benchmark for innovation. And you might ask why. Simply put, 3M has mainstreamed innovation. Its commitment means innovation comes from anywhere and everywhere in the organization. No one person or department is given the responsibility for innovation; instead, it is everyone's job or responsibility.

Similar to the mystique of Disney for customer service, 3M opens its doors for visitors to tour its facilities and learn the "key" to innovation (like Disney's Keys to Excellence program). And in both cases, those visitors who seek a cookbook recipe leave disappointed. These companies achieved their success with more than one policy or one process. It is about a total alignment of organizational resources and a commitment to fostering a culture supporting this alignment. Our job, then, is to understand enough of the elements and resources to be able to integrate them into our own unique innovative culture. It is about creating the DNA that works for us!

Tony Hsieh, chief executive of Zappos, is clear in his message that "culture is priority one!"—even using this as the title of his YouTube video.[19] Ross Tartell, technical training and communication manager, GE Capital Real Estate-North America, tells it straight: "Culture is everything! Culture trumps strategy!"[20] Adaptive cultures provide their companies with a competitive advantage in outperforming the competition. Corporate culture is identified as the number one driver of innovation.

HR's Role: The Heavy Lifting

As HR professionals, we are responsible in large part for nurturing the culture of our organizations. As we shift to more strategic work, this challenge becomes even more critical. Working at the epicenter of our organizations, we have the opportunity to shape the components essential to effective execution in building this culture.

Remember: One size does not fit all. What matters in developing an innovative culture is how we put the pieces together in our own organizations.

As the guardians of the corporate culture, the responsibility for guiding an innovative culture rests largely with HR professionals. This, then, requires that we have a clear understanding of the elements generally possessed by a culture of innovation. Using this as our foundation, we can better perform the recruitment process to ensure creative talent feeds this culture. And once hired, this talent must be developed, managed, and recognized to further prime the pump of innovation throughout the organization. Leadership also has a clear role that is crucial in fostering the innovative culture. As HR professionals, this partnership is essential with top management as we nurture the culture of innovation.

Spotlight on Innovation

"Innovation and creativity in the workplace have become increasingly important determinants of organizational performance, success, and longer-term survival."[21]

Though the lives of miners may not be at stake, our organizations' very survival may be at risk. We must see that limitless possibilities (albeit different) exist with innovation! Our commitment to innovation, then, as HR professionals is not "added on" to our existing responsibilities. Instead, we must see it as integrated into all we do—from recruiting to training and development. It must become a part of our own language and the processes in which we engage. Stanleigh advised that "HR leaders need to understand the critical importance of innovation today and how to contribute to constantly improving their skills and creating a culture of innovation. This will enable your organization to differentiate itself. These are a part of the role of HR."[22]

Limitless possibilities can be generated as diverse people come together—in the *right* way. It is not just about throwing different people together, but rather true integration is needed for success. The subsequent chapters will provide insights into how we can generate these opportunities by creating a new infrastructure and developing innovation competencies to nurture our culture.

Figure 1.2. The 9 Dot Solution

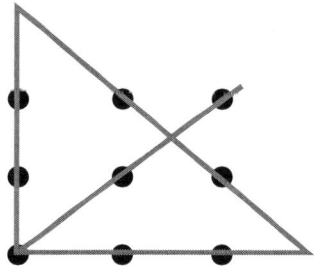

Chapter 2.

Understanding the Foundation of an Innovative Culture: Making It Work!

Author and consultant Scott Edinger stated that "creating a culture where innovation thrives ... can become a source of competitive advantage."[1] A 2009 study highlighted the importance of culture in the innovative organization. In predicting which companies were most likely to innovate, corporate culture was cited as "the most important driver."[2]

Structural changes to foster innovation are discussed in this chapter with some of the characteristics that build a foundation for the more innovative organizations. These characteristics include embracing an unorthodox perspective on failure and mistakes, creating a free flow of communication, promoting a collaborative environment, encouraging time for reflection with scheduling unstructured time, and providing the resources necessary for employees to be innovative. The essence of an innovative culture must come from the top of the organization and trickle down throughout the firm—permeating all levels. Trust has become the foundation of the characteristics of an innovative culture; therefore, the innovative culture goes hand in hand with a culture of trust.

Contrary to popular belief, innovation is not just for the small entrepreneurial venture or the technology firm. It is important for the older, established companies as well—if they want to thrive. And perhaps in today's dynamic environment, a focus on innovation is even more meaningful for larger organizations where the trap of bureaucracy looms. A sampling of best practices is presented throughout the chapter. Examples from companies such as 3M and IDEO (known for its innovative cultures) are examined with particular attention to the pervasiveness of a commitment to innovation. 3M's classic Post-it Note development is discussed from the "big firm" perspective in providing guidance for other companies.

Spotlight on Innovation
"Innovative success happens in volume."[3]

The Power of Failure: Revving the Engine of Innovation

Perhaps the greatest overlooked source of energy driving the engines of our innovative cultures may lie in our failures. It is time to reframe the way we think about failure. The number one myth about failure is that it is bad. Failure, however, is not necessarily or inherently bad. Recognizing the power of failure provides an opportunity!

In environments of uncertainty, success is more likely to occur as we increase the number of attempts made. We must widen the funnel for attempts! This means, however, that we will produce more failures. And these failures cannot be punished if we are going to encourage people to continue these attempts and experiments.

Babe Ruth, often regarded as one of the best baseball legends in the history of the game, racked up a record 714 career home runs. Yet this success did not happen overnight. Ruth knew the power of failure. He mined his failures—otherwise known as his strikes—for corrective action and knowledge to improve his game. He knew his strikes (his "misses") served as one more step toward the home runs he sought.

When asked about their organizations' ability to manage and learn from failure, the overwhelming majority of executives admit ineffectiveness at best. The traditional approach has been to explain away failures in a positive light or to ignore them altogether. Facing failure and even embracing it for what it is—an opportunity to learn—have been viewed as risky to a manager's career. Failure, however, can be mined for valuable lessons if we become more comfortable with the potential risk. The organizational culture can reinforce this comfort by cultivating an environment of trust where it is safe to fail.

An environment where people feel safe making intelligent mistakes is essential. For example, philosophy of former chairman of the board William L. McKnight at 3M has been widely shared. According to Ernest Gundling, "Management that is destructively critical when mistakes are made kills initiative, and it is essential that we have many people with initiative if we are to continue to grow."[4] 3M's persistence in the face of repeated failure certainly paid off. The breakthrough for Scotchlite sheeting was reported to have taken 12 years!

The importance of failure—and learning from it—is also understood by many venture capitalists. Overall, they are more likely to invest in an entrepreneur who has already experienced failure. (But of course, we would imagine they would want to see evidence that the entrepreneur learned from the failure.) In a *Harvard Business Review* article, professor Daniel Isenberg referred to the "cult of failure." Isenberg shared how successful entrepreneurs tend to view failure in a healthy way. With this perspective, they are not paralyzed by the fear of failure. However, they do recognize the need to fail fast—and early.[5] The use of prototypes in our organizations has helped us be fast in innovating. This provides the time (and the opportunity) to learn and move forward.

Spotlight on Innovation

"The courage to confront our own and others' imperfections is crucial to solving the apparent contradiction of wanting neither to discourage the reporting of problems nor to create an environment in which anything goes. This means that managers must ask employees to be brave and speak up—and must not respond by expressing anger or strong disapproval of what may at first appear to be incompetence."[6]

We certainly need a sense of urgency to solve the problems at hand today. The window of opportunity is not open for a limitless amount of time. This provides additional incentive then for us to quickly recognize failure so we can move on and keep the momentum going. Perhaps today's updated, more appropriate mantra for being a successful innovator is to fail often and fail fast!

It is important to be decisive. The idea is to avoid wasting additional resources. Ironically enough, identifying the valuable lessons from failure when it is fast more easily helps us connect the dots between our actions and the unfavorable outcomes. When too much time passes, we may find it more difficult to connect those dots and learn our lessons. Isenberg has another perspective on why we need speed in failing; he suggested that "rapid failure functions like the draft of a chimney: The fast exit of failures sucks in new entrants."[7] Furthermore, sometimes moving ahead quickly on a small scale lets us avoid the greater expense of a possible larger failure. Pilots and prototypes are good alternatives for organizations to explore to fail faster (and on a smaller scale). As hard as it is, the tendency to perfect that prototype may need to be traded for speed.

Safe to Fail

The first step toward learning from failure is to create an organizational climate that does not punish failure and instead encourages intelligent failure. The term "instructive failures" has been used to reinforce this learning element and suggests that these failures should generate new knowledge.

Organizational leadership sometimes hesitates to provide this safe environment to fail due to misunderstanding the results. An unfounded (though popular) myth is that if, as managers, we create an understanding environment around mistakes, it may result in a loose environment (or perhaps even worse—chaos) where even more mistakes occur. As part of an innovative culture, this view should be reevaluated. Perhaps, instead, we need to turn this thinking upside down. Harvard Business School professor Amy C. Edmondson suggested, "Those that catch, correct, and learn from failure before others do will succeed."[8]

Spotlight on Innovation
"If your organization can adopt the concept of intelligent failure, it will become more agile, better at risk taking, and more adept at organizational learning."[9]

If properly framed, we can use failure and its lessons as a path to success. However, if our organizational culture allows us to play the blame game, the lessons of failure are usually not recognized— and certainly not shared in the organization. Therefore, we need a safe environment where people are comfortable talking about their failures (rather than covering them up). The culture must create a psychologically safe environment where people can fail—and live to tell about it. Our managers and leaders are responsible for reinforcing a culture that does not support the blame game. To do so, the focus must be shifted; it should not be on who did it, but instead on what happened. This, then, is the systematic analysis that generates the new knowledge of the intelligent failure.

Julianne M. Morath, former chief operating officer at Children's Hospitals and Clinics of Minnesota, built just that type of culture. The emphasis is on communicating the types of failures most likely to be experienced and why sharing these failures is valuable. The blame game is not played. Rather, the focus is truly on learning— critical in a hospital environment that manages life and death situations on a regular basis.

But we should be clear: There is a caveat to embracing failure. We must think in terms of degrees or types of failure. We certainly do not want to encourage just *any* failure. All failure is not equal. It is critical to differentiate failures that are not acceptable and to take action to hold those accountable. We must explain those violations and consequences to others. Failures that can be prevented in routine operations are considered unacceptable. And limits must be identified. UPS identified its off-limits zone for failure as those areas directly affecting the customer. For example, failures that negatively affect the customer experience are not considered acceptable failures at UPS.

Learning from Failure

The leadership of organizations must be the role models. They can give others permission to talk about failure and even encourage these discussions. These might be considered "mini-postmortems." These reflections are vital to understanding what went wrong—and how to avoid repeating it. Therefore, these lessons must be widely shared. Some have gone as far as to suggest we need to publicly and widely celebrate our failures!

Chairman, CEO, and president of Procter & Gamble (P&G), A. G. Lafley suggested that leaders stop growing when they are no longer learning. Under Lafley's leadership at P&G, the company "studied ... failure in detail." Lafley recognized the importance of learning from mistakes (and communicating those lessons) or being doomed to repeat them. Karen Dillon of the *Harvard Business Review* shared that Lafley even viewed "failure as a gift."[10] But of course, we can determine what the gift is only when it is opened!

Spotlight on Innovation

"If it ain't broke, experiment."[11]

Essential institutional learning results only from an environment that is courageous and open to improving. This environment does not shoot the messenger, but rather embraces the learning.

And in fact, we must solicit feedback and participation around our failures. And we must encourage people to observe and analyze failures everywhere. We might even consider providing rewards to reinforce this behavior.

It is easy enough for organizations to talk a good game about embracing failure. It is much more difficult (and less common) for companies to actually walk the talk. We often fall into our old traps. To avoid the admission of failure we sometimes change (or recalibrate) our definition of failure. We want to beware of recalibrating deviations we consider acceptable because this redefines failure and covers up opportunities for learning. Before automatically redefining failure, we want to seek out the cause (versus simply correct the symptoms).

In the analysis of failure, we must drill down to the root causes. Otherwise, we miss the opportunity to capture the right lessons. Unfortunately, connecting the dots about our failures—that is, to assign causes—can be challenging. It is necessary to understand the series of small events that may have led to a failure. Projects usually have multiple evaluation points that we can use to check our status. Clearly identifying what success (and failure) looks like at the start will help us in our analysis. We recognize, however, that deconstructing our failures is difficult emotionally. We have a tendency to want to move on quickly rather than properly mine the failure for all the organizational gold.

The "constructive stumbling" embraced by 3M is the idea of being in perpetual motion. To do so, management must support employees in the exploration of a variety of ideas. When failures occur, takeaways must be gleaned, and employees must be encouraged to continuously seek out new solutions and move forward. W. L. Gore & Associates, a U.S. manufacturing company, conducts post-mortems on failed projects to learn—with no punishment. We want to fail often, fail fast, and then bounce back to move ahead. This is a measure of our organizations' ability to be resilient!

The Flip Side of Failure: Don't Forget to Analyze the Successes

We must exert the same rigor to examine the causes of our successes as well as of our failures. As difficult as we find analyzing our failures, analyzing our successes proves even more challenging. Harvard Business School professors Francesca Gino and Gary P. Pisano highlighted the irony by making the point that "nothing inflates confidence like success."[12] Yet success and its accompanying confidence make us less likely to reflect on what happened. There is a tendency when successful results are achieved to assume that the processes leading up to that success are sound. This assumption is dangerous at best and may even provide a missed opportunity for learning.

Unfortunately, when we succeed, we assume we should keep doing the same thing—rather than evaluate our assumptions and build our knowledge—that is, learning. In and of itself this is a dangerous assumption. Though it may seem counterintuitive, we need to devote time and energy to carefully analyze and study our successes. A failure to do so may ultimately result in poor decisions going forward.

With this overconfidence generated by our success, we tend to value our own opinions more than the opinions of others. This contributes to erecting barriers to the effective communication of those bringing bad news. And therefore, we do not hear the information we need. We engage in selective hearing whereby we hear only what we *want* to hear and information that confirms what we *think* we already know. And thus, the seeds of our downfall are sown.

Spotlight on Innovation

"Failure and success are on equal footing and both trigger further investigation that helps us revise our assumptions, models, and theories."[13]

Furthermore, confidence leads to more confidence—and perhaps the overconfidence bias whereby our judgment is clouded as we become too full of our collective selves. The erroneous assumption is that if we were successful, we do not need to examine what happened. However, failure to examine success can lead to an organization's decline. This was the case with the Ducati racing team, which helps explain its fall from a record of Formula One success.

The fundamental attribution error clouds our view of success. We tend to attribute the cause of our success to internal factors (for example, our strategy). We also tend to underestimate how many factors (many of which are beyond our control) contribute to our success. We assume our success resulted from what we did and what we knew. And rarely do we see it as a "lucky hit"—or an external factor. So again, we have missed an opportunity for learning.

Our vision is blurred and perhaps time-bound when probing our successes. We tend to forget the long lead times between taking action (or adopting strategies) and the results we achieved. As a result, we look at what we are doing now and not what we *were* doing that led to those results.

The U.S. military's after-action reviews provide an example of a best practice for companies to adopt. Both successes and failures are examined. The military recognizes the value in analyzing both what went well (and should be repeated) and what went wrong (and should not be repeated). On a smaller scale, the performance appraisal in some organizations exemplifies this approach. According to Gino and Pisano, "Understanding the reasons behind the good performance of successful employees may bring to light important lessons for others."[14] Instead of focusing only on the areas in need of improvement for poorer performing employees, we might consider sharing more of the lessons from our top performers. Now there is an innovative approach. (Innovative HR practices are further discussed in Chapter 7.)

The Near Misses: An Opportunity Not to Be Missed!

Near misses should be given as much discussion and attention as large scale, flat-out failures. There are still lessons to be captured, yet we have a tendency to let these slip by us. We view them as unimportant, perhaps even seeing near misses as "lucky saves." As a result, we often quickly move on without considering what happened and are happy to put the near miss behind us. Unfortunately, we have overlooked an important opportunity—a preventive maintenance of sorts. A thorough postmortem on a huge failure often reveals a number of near misses that were considered uneventful. We simply ignored them and therefore lost a valuable opportunity to learn.

When we address near misses, we often view them as anomalies rather than as potential failures. When we accept anomalies, we become more and more comfortable with these deviations and forego the opportunity to learn from them. A better approach might be to embrace our near misses as failures and use them for the lessons to be learned to prevent future failures on a larger scale. Professors Catherine H. Tinsley, Robin L. Dillon, and Peter M. Madsen reported that "research on workplace safety, for example, estimates that for every 1,000 near misses, one accident results in a serious injury or fatality, at least 10 smaller accidents cause minor injuries, and 30 cause property damage but no injury. Identifying near misses and addressing the latent errors that give rise to them can head off the even the [sic] more mundane problems that distract organizations and sap their resources."[16]

The innovative culture rewards and encourages discussing—and even highlighting—the near misses. This approach is recognized as an investment in our future success and an opportunity for learning.

Communicating a Commitment to Innovation: Beyond Just What We Say

Just as the culture of each organization is unique, the approach to innovation in each organization is different. There is no one right way or one best path. There are, in fact, multiple paths for our organizations to achieve innovative cultures. However, a key element identified as a differentiator for more effective innovative companies includes a commitment to innovation by top management. And this commitment must be clearly communicated throughout the organization. Samsung is well known for its management program launched in 1993. This initiative altered the culture of the firm to focus on innovation—starting a period of exceptional company growth. This management program was a mechanism by which Samsung widely communicated the importance of innovation.

Communication is essential in fostering a culture of innovation. And top management plays a critical role here. It must clearly and continuously communicate that innovation matters. This is not a once-a-year announcement or a corporate value written in a marketing piece or distributed on a laminated wallet card. It is a living, ongoing commitment that is discussed routinely and is broadly known throughout the organization. Top management's role is to keep innovation on our agenda all the time.

3M understands the importance of clearly communicating a continuous commitment to innovation. Every year it specifically targets a portion of its annual sales that are expected to come from products that were launched within the past four years. This innovation target becomes a part of the company's strategic plan and is widely discussed throughout the organization.

Simply describing innovation or identifying innovative target goals is not enough. This commitment to innovation must be communicated in what we do as well. The nonverbal elements of communication become just as powerful (and perhaps more powerful) than what company leadership says. When Intuit created the position of vice president of innovation, it sent a clear signal of its commitment. Budgeting for innovation signals its importance, and setting aside discretionary funds for projects that may arise in the future speaks volumes. Providing rewards and recognition for innovation sends the message about what is valued in the organization.

Communicating a commitment to innovation means that we must embrace the culture of trust and transparency across the organization.

Spotlight on Innovation

"Innovation shouldn't be walled off because it is supposed to be everywhere."[17]

Encouraging Open, Transparent Communication

Fostering a culture of innovation requires open, transparent communication. Information must be shared upward, downward, and laterally. Unfortunately, most organizations tend to filter the information that flows upward. A psychologically safe environment allows more of the "bad" news to be shared upward, but a culture of fear increases the probability that the information needed will not be communicated upward. Instead, the tendency in this environment is to engage in cover-ups, finger-pointing, and avoidance, resulting in top management's being unaware and sometimes blindsided.

A culture of innovation must specifically address this and combat it with an openness and emphasis on transparency. We must encourage people to communicate openly, sharing their failures versus hiding them or allowing them to fester. And we must engender an openness to share ideas and collaborate. It is vital to openly share ideas on *all* fronts, including on what is working and what is not working, what we know and what we do not know. Knowing what we *do not* know is significant, and we must let others know this. That is, the environment must embrace the fact that it is okay not to know everything.

Eli Lilly encourages open communication—even about mistakes. For over 20 years, the company has held "failure parties" for intelligent failures. Part of this party is having people transferred to new projects. This is a public way to acknowledge the mistake, to communicate that there is life (and "good" life) after intelligent failures, and then to move on.

Downward communication must also be open to foster innovation. If information is controlled and hoarded by top management, the innovation of employees throughout the organization is stifled. Open, transparent communication, then, is a foundation for a collaborative environment.

Building a Collaborative Environment

An innovative culture requires a forum for collaborating. Innovation does not happen in a vacuum. Seldom do our innovative ideas develop while we are sitting alone in a dark cubicle. Instead, collaboration often provides the nurturing environment to grow our ideas and to help grow the ideas of others. Building relationships and supporters for our ideas across the organization is possible only with open communication. Even more, encouraging collaboration helps elicit different perspectives—critical to the creative process. Collaboration requires that we keep others involved and informed.

Spotlight on Innovation

"A culture of innovation grows because everyone can play. . . . Every employee can be a potential idea scout and project initiator."[18]

We want to encourage people to pool ideas and even borrow ideas from other departments in our own organizations and other companies—even other industries. Encouraging collaborative thinking includes looking outside the organization (externally) for ideas. It is not to copy those ideas, but to use them in novel ways in our own organizations.

The creation of cross-functional teams provides a forum for bringing diverse perspectives together. The speed with which new knowledge becomes available has grown faster and faster, which has made staying current in our fields harder and harder. Collaboration allows us to pool our knowledge and build on the knowledge and ideas of others—to stand on their shoulders, so to speak. But of course, it is the foundation of open, honest communication that encourages this collaborative environment. We must feel safe, not threatened, in sharing.

Furthermore, encouraging employees to establish strong bonds is essential for successful collaborations. This produces a culture based on respect and trust where individuals not only tolerate differing opinions but embrace divergent thinking. It is an environment where individuals feel safe to disagree with others and openly deal with conflict. Individuals must feel that they have permission to challenge the status quo—at a basic level. Each employee feels comfortable being that lone individual to take an initial stand. Organizations must then structure processes to ensure the exchange of ideas and foster more open communication.

An example of building an environment for communicating collaboratively comes from IBM on quite a large scale. Chairman and founder of Xyntéo Ltd. Osvald M. Bjelland and Professor Robert Chapman Wood described a "massively parallel conference online" known as "Jam" whereby IBM employees are able to connect online through bulletin boards.[19] Answers to questions are posted, and new ideas are often generated and shared. The company recognized that ideas are plentiful and come from many individuals throughout the organization. It is almost like a mega brainstorming/war room session held virtually in an online forum.

Whatever the technology we choose to use or leverage and however we decide to connect people (including virtual teams), the focus must be on open communication and the free exchange of ideas spanning a wide range of boundaries. A key lesson learned from IBM is to include everyone in our open communication—rather than include only a select few "innovative types." After all, innovation is everyone's job!

Spotlight on Innovation

"Nowadays, it's just not possible for individuals, no matter how expert, to develop important innovations all by themselves. The chances of individual parts, developed separately, coming together into meaningful, functional wholes—a new product, feature film or rescue operation—without intense communication across boundaries are exceedingly low."[20]

How Flexible Are We?

Fostering an innovative culture requires that we harness the potential of individuals. The very creative personality types and innovative behaviors that are essential to developing this culture require that we support divergent thinking, often not an easy task. This support extends to helping individuals pursue their ideas—though often different—and to reframe our view of failure and risk. And finally, we must provide a forum for them to build collaborative relationships in the organization. Without a feeling of connectedness and engagement, individuals do not feel comfortable, and innovation is not achieved. Supporting collaboration requires flexibility in both our structures and our processes.

To encourage collaboration, then, we may want to consider more fluid organizational structures. The traditional silo approach to organizational design offers inherently less collaboration and information sharing. Too much hierarchy can stifle creativity. In fact, overstructuring most things in the organization will stifle creativity. Narrow job descriptions have been proclaimed an innovation killer because we overstructure, overdefine, and thereby limit and constrain the job incumbent. There is no room left to think creatively.

Some child psychologists have voiced concern with the overscheduling of young children. Millennials are considered the most overscheduled generation in history. And we also know that they value constant feedback on how they are doing. Is there a connection? Overstructuring lives may develop children and young adults who are not creative when left to their own devices. Perhaps a little less structure can be a good thing.

We need to build in some spontaneity. Flexibility and freedom are key ingredients to an innovative culture. We must consider providing employees with a way to collaborate and interact (and the time to do so). Building in flexibility includes allowing for time to think and the freedom to create. It is difficult to be innovative if every minute is accounted for with scheduled tasks. 3M's "15 percent rule"[21] provides employees with 1 hour and 12 minutes during each 8-hour day to pursue their own projects. Daniel Pink, in his book *Drive*, refers to "small islands of autonomy" for noncommissioned work.[22] Twitter's 2010 "Hack Week" was one designated week of noncommissioned work that allowed employees to work on projects outside of their normal work.

In general, then, we need to guard against the tendency to overmanage. In fact, micromanagement has been described as the antidote to creativity. A controlled environment is not conducive to cultivating innovative ideas. An openness to change may require a willingness to let go and even shake up the status quo. However, placing less emphasis on hierarchy and foregoing rigid structures does not mean we have to allow chaos. We *can* have a disciplined approach to innovation—but with less structure and hierarchy. Providing clear, meaningful goals is imperative. Then we need to step back, fight the tendency to micromanage, and simply let our workforces use the skills we hired them to use. (There is more on giving up control in Chapter 6.)

Spotlight on Innovation

"When people come together outside the confines of the traditional office or conference room, there is a more natural open dialog."[23]

Members of an innovative culture recognize that accepting or even embracing change is simply not enough. Today's organizational members at all levels must be proactive in creating change through innovation. These individuals, then, must be participative decision-makers with clear goals and the autonomy to pursue them in novel ways. A nonbureaucratic structure with cross-functional teams fosters curiosity among its workforce. Being flexible and adaptable enables us to create win-win situations. Google uses project teams with extremely large spans of control—with little control and hierarchy.

Tight controls have the potential to create huge barriers to innovation. And yet we have a tendency to use a one-size-fits-all approach to much of what we do in our organizations. Rosabeth Moss Kanter of the Harvard Business School pointed out that "the same planning, budgeting, and reviews applied to existing businesses" may hinder our ability to be innovative in new businesses.[24] It is often appropriate to measure new businesses with different metrics. BankBoston learned the value of moving away from uniform financial metrics. As BankBoston, it successfully created First Community

Bank and evaluated this business based on new metrics to give it the flexibility it needed to grow and prosper.

Highly structured performance standards used in performance appraisals should raise red flags regarding the role of rigid processes in stifling creativity. With our heads down to achieve the standards set, we may miss valuable opportunities to be creative. There is no chance to veer off on another course to try something else as we keep our eye on the singular target of the prescribed standard. Freedom, flexibility, and less control are critical for innovation to allow individuals and teams to try new things.

Mainstreaming Innovation

Some organizations have chosen to isolate their smaller innovative businesses to avoid tainting by the bureaucracy and hierarchy of the bigger organization. An unexpected lesson was learned by General Motors, however. The company successfully launched Saturn in a separate business unit to allow it to be innovative and free from the red tape and bureaucracy of the organization. Then when Saturn was brought back under the GM umbrella, instead of bringing back innovative ideas, Saturn became more like GM—losing some of its innovative edge.

This concept of separate innovative structures within our organizations has the potential to raise yet another red flag. We must be careful of creating two cultures: one where all the fun resides and one where the "business of the business" is conducted. One is required to play by rigid rules, and one is perceived to have few or seemingly no rules. This is *not* today's model. We want innovation to *permeate* the entire firm—not just a select few departments.

As we have seen, communicating that innovation is important across the organization is an essential component of fostering an innovative culture. But that is only the start. If innovation is truly mainstreamed, it is part of all we do. We must ask if we walk the talk. Do we then provide the necessary resources to make innovation happen? Do our organizational values provide the framework for our decisions and even determine how we allocate our resources and rewards? We need to focus on what leaders do versus on what they say is valued. That is, do we truly support an innovative culture throughout the organization, making it the responsibility of everyone?

Spotlight on Innovation

"Work environments have an impact on creativity by affecting components that contribute to creativity, which represent a basic source for organizational innovation."[25]

Trust: The Glue That Binds

The importance of trust cannot be emphasized enough in fostering an innovative culture. Creating a safe environment to fail cannot happen without trust. The transparency required for open communication cannot occur without trust. Collaboration among individuals, teams, and external stakeholders cannot be effectively cultivated without trust. And flexibility and freedom cannot be provided without trust. Trust is the glue that binds all the elements of culture together—making them possible. A culture of trust, though, takes time and an ongoing commitment. One small misstep can easily break the trust we have so carefully cultivated.

Cultivating an environment of trust requires that we do what we say we will do. As innovators are hired, a culture of innovation will engage them and provide them with the environment to do what they have been hired to do—innovate! In this culture, members of management walk the talk providing role models for innovation both in what they do and say. Innovation is rewarded and recognized with a widely communicated process for generating and receiving innovative ideas. Resources (both financial and nonfinancial) are provided to support innovation throughout the organization. Leadership development programs and training emphasize the importance of innovation and skills needed to effectively innovate.

This trust flows both ways. Management trusts employees enough to empower them and to provide them with freedom, and employees trust management to run interference to remove any blocks to innovation. Management trusts that employees can be given broad stretch goals, and employees trust management to let them meet the goals in the way the employees determine is best and reasonable.

An innovative culture can be a company's source of competitive advantage. Just ask 3M. Yet reaping the benefits of an innovative culture requires more than just adding an element here or an element there. For example, we cannot focus on recruiting innovators and then throw them into a highly controlled, bureaucratic structure where they will wither on the vine.

A culture of innovation recruits and develops people who are passionate about innovation and are inquisitive, independent thinkers and lifelong learners. Top management steadfastly embraces innovation and the organizational processes required to engage our innovators and to foster innovation throughout the organization—including management's own processes. An environment of empowerment contributes to the collaboration of cross-functional teams and provides autonomy for innovation to thrive.

Corporate culture is the most important driver of innovation. We certainly need the resources and people, but they are only effective if we develop the culture to effectively deliver innovation.

It is only with a foundation of trust that this can be achieved. The subsequent chapters address the details of additional elements required in fostering an innovative culture.

Spotlight on Innovation

"Finding innovation is almost a sacred quest for the solution that will create growth, and open new eras of prosperity and well-being."[26]

Chapter 3

Hiring for Innovation: Feeding Our Culture

A major barrier to recruitment today is thinking we can do without creative talent if we are not a high-tech firm. David Burkus, author of the article "10 Practices That Drive Innovation," warned that "the war for talent is slowing [sic] shifting its focus from quantitative minds to creative ones."[1] As the guardian of the corporate culture, HR plays a critical role in the recruitment and selection of individuals who "fit" a culture of innovation. These individuals are essential elements when fostering an innovative culture. Characteristics of innovative individuals are explored in this chapter. These include persistence, a natural curiosity, adaptability, strategic thinking, and resilience.

Hiring for innovation requires a long-term focus because we are making an investment in recruiting people for the future—not just for today's job. HR's ability to influence the hiring process with others throughout the organization is central to the innovative culture. We must make a concerted effort to hire creative individuals to advance our innovative culture. The use of the interview as a marketing tool, adopting a true commitment to hiring a diverse workforce, and hiring for fit are explored in this chapter. In addition, we will explore how the well-publicized commitment by Zappos to hiring for cultural fit provides a best practice.

Spotlight on Innovation

"Select for leadership and interpersonal skills, and surround innovators with a supportive culture of collaboration."[2]

Marketing Our Innovative Culture: A Lesson from Google

When benchmarking best practices for hiring creative talent, companies often refer to Google, and then on closer examination, many immediately dismiss that model. Most executives are more comfortable waging the war for talent by throwing money at it. Google, however, relies more on the influence of top management and offering freedom (over money). That is, it lures recruits into the organization by providing the freedom and autonomy that allows our creative types to thrive.

It is often more difficult for leadership to give up control than money. Furthermore, top leaders must change their perspective to think in terms of the innovative ideas coming from their employees, rather than of their employees looking to them for ideas. Recruiting the right candidates takes on even more importance when we recognize they are the ones most likely to add value to the organization—not leadership!

A demonstrated commitment to innovation can also be a great marketing hook to hire talent. An essential part of our branding, it can serve as a magnet for innovative individuals. Innovators and creative individuals are drawn to environments where they can thrive and do their thing. These are the individuals who sustain the culture and help draw others in—thereby creating a self-perpetuating culture.

The Three-Pronged Approach to Hiring: Hiring for Culture

Recruitment is key since people are key. Our people are the critical element of innovation. The right employees can feed the innovative process, and the wrong employees can hinder or even destroy the innovative process. In today's competitive business arena where talent management is a serious challenge for organizations, simply hiring the right person to fit the job is not enough. To make the best decision requires that the person also fit the organization's culture. Using the three-pronged approach (which addresses the person/job/

culture fit) increases the probability of selecting individuals more likely to be effective performers within our organizations, more likely to be engaged, and more likely to stay longer.

Zappos has been highlighted as an organization that effectively hires for cultural fit. Executive candidates who do not contribute their time on the phones delivering the famous Zappos "wow" of customer service are deemed culture misfits and are not hired. Tony Hsieh, CEO of Zappos, stands by his approach of firing for culture as well. He further advocates paying people to leave! The Zappos example stresses the importance of protecting our culture—albeit in somewhat different ways.

Spotlight on Innovation

"By mindfully seeking architects and designers from different backgrounds, one architecture firm proves that variety in staffing results in more creativity, innovation and better problem solving."[3]

Disney's hiring process has used the approach of letting candidates self-select out of the process. Group interviews enable job applicants to interact with others to assess their own cultural fit. During the hiring process, we must likewise attempt to screen out those who are not a good fit.

The Zappos website is clear in its message concerning its culture:

> At Zappos, we give you the creative freedom and autonomy to follow your passions in a way that suits you best. As a company that practices Holacracy, we fight the strains of bureaucracy by replacing it with a new self-organizing system designed to focus around fulfilling work without office politics. We also promote a team and family spirit that fosters camaraderie across the company through cool, collaborative spaces.[4]

This approach provides an opportunity up front for those potential applicants who crave bureaucracy, politics, and rigid rules to opt out.

Diversity in Recruitment

We must consider diversity when recruiting for the innovative culture. The seeds of innovation are better watered and nurtured with different perspectives. Although there is often a natural tendency to surround ourselves with others who are similar to us, this is counterproductive. Creative ideas and innovative solutions are more likely to be generated when we bring diverse people together and provide them with an environment that nurtures this approach. This diversity includes background, personality, and perspective. The innovative culture embraces diversity beyond basic demographics to include work style and work experience.

Part of this diversity involves being aware of the decision to recruit from within or outside the organization. It is important to bring invigorating individuals (and ideas) into the organization. An overemphasis on hiring from within stems the flow of new blood. A balanced approach provides a win-win to refresh our human capital and its perspectives. Hiring from within may also involve transferring individuals among functions—further providing new perspectives and a cross-pollination of ideas.

The co-founders of EiQ, Terry M. Farmer and Xavier Butte, suggested that innovators "connect to a diversity of opinions, perspectives, and ideas through people, data, experiences and analogies."[5] Innovation requires the integration of information from different sources, so it only makes sense that the integration of people with diverse information and knowledge bases would contribute to the innovation process. We want to seek out individuals, then, with broad interests who are comfortable with diversity and who are open to different perspectives.

In an interview with Gordon Carrier, principal at a design company, Luke Siuty reinforced Carrier's commitment to multicultural diversity in his creative organization. Carrier believes "diversity catalyzes innovation" and strengthens the organization—serving as a critical asset to the organization. He further said that "innovation ... is about points of view, and diversity by definition is point of view. ... The fact that multiple voices are weighing in on design problems ... gives us an innate advantage."[6]

Although an inability to question ourselves provides a roadblock to innovation, we experience even greater barriers when we close our minds to other opinions and dissenting viewpoints. Surrounding ourselves with those who think the same as we do is not creative and leads to a unilateral lens. Different perspectives are critical! We must fight the urge to surround ourselves with others just like us. Reading more widely and interacting with more diverse individuals helps us open up our own thinking. As we expand our openness, we can bounce ideas off others—and even solicit the input and participation of others to build on our knowledge.

Expert Knowledge: Simply the Buy-In

Organizations must first and foremost hire individuals who are knowledgeable. Yet this is simply the minimum level of buy-in needed. Just as a gambler must have a minimum amount of money to buy into a poker game, so it is with hiring. Though we seek many characteristics, the minimum is expert knowledge.

A strong knowledge base is essential to an innovative culture, but it can also be a double-edged sword. Specialists with depth of knowledge must guard against tunnel vision and the idea that they already know it all. This knowledge of what is known with certainty can blind individuals to alternatives and additional approaches. Formal education can sometimes drive out creativity if we are not self-aware. Education can arm individuals with a false sense of confidence in having all the right answers. Individuals who wed formal education with a natural curiosity are more likely to be innovators.

A balance is needed between in-depth, specific knowledge and knowing a little about a lot of areas. This provides the foundation for making the connections between seemingly unrelated subjects that is so critical to the innovative process. Individuals who are well read and who tend to read broadly are in a better position to integrate information from various sources.

A specialist with subject area expertise who is also somewhat of a generalist provides the best of both worlds. The generalist's perspective enables individuals to better see the big picture. As Harvard Business School professor Amy C. Edmondson suggested, it is also possible to bring a generalist and specialist together to achieve the best of both worlds—combining their distinct perspectives (which is exactly why we want to hire for diversity).[7]

Making connections between seemingly unrelated things is important to the creative process. This ability to recognize patterns and make connections is known as lateral thinking. Those individuals who have broad interdisciplinary knowledge tend to be better able to do this. They are able to integrate knowledge from multiple areas and challenge more conventional thinking. They are also able to bring the expertise of others together to generate a wide array of ideas and then connect them in unconventional ways.

This is exactly what Steve Jobs did. He brought ideas together from different areas in a novel way. He launched Pixar with the movie *Toy Story*. Jobs combined the animation software from George Lucas with storytelling. For another example, combining opposing thoughts made Picasso famous as his art became known for the combination of a variety of techniques and ideas.

Spotlight on Innovation

"Creative individuals have the ability to see problems in unique ways in order to produce solutions that are equally unique."[8]

Strategic Thinkers: Apply Within

A strategic approach to thinking is considered one of the major components of a high-performance organization. Yet strategic thinking has been identified as a deficiency in many organizations. The good news is that as we attempt to address this deficiency, we are concurrently helping fuel our innovative engine. Innovators tend to be more strategic in their thinking. Understanding the role of the big picture provides an essential fundamental building block for strategic thinking and for effective innovation.

The big picture might be compared to a helicopter view—or the view from 5,000 feet. A more tactical view would be the perspective from the ground. This big picture might be acquired by stepping outside our function—literally! This aerial view enables us to see what we miss from the ground while in the midst of the everyday transactions. The ability to see the big picture means that professionals are skilled in understanding how all the parts of the organization interact and fit together. This is the view from 5,000 feet. Furthermore, it means that what we do is linked to the organization's mission and objectives.

Strategic thinkers engage in synthesis. That is, they gather information from multiple sources. Problem-solving is improved by using multiple perspectives to see things differently. This can be accomplished by using both divergent and convergent thinking. The two approaches to thinking have been compared to the use of a wide-angle lens and telephoto lens. The telephoto lens enables the photographer to zoom in, while the wide angle lens captures the entire view. Just as a skilled photographer would recommend a zoom lens as more ideal (to enable both functions), the strategic thinker likewise benefits from this ability to zoom in on one detail or zoom out to see the broader view.

Strategic thinkers take a systemwide approach to the business. That is, they recognize the interconnectedness of all the functions of business and acknowledge the ripple effect. When a change is made in one function of the organization, it will most likely affect other areas of the organization. Working without strategic thinking has been described as an organizational lobotomy[9] creating the effect that one hand really does not know what the other is doing.

Spotlight on Innovation

"Innovating, whether in product development or problem solving, means working without a blueprint."[10]

The tendency to fall into a silo mentality must be avoided. This places blinders on individuals who then see the problems (and the organization) through the lens of only their departments and functions. That is, accounting sees problems from only an accounting perspective without regard to marketing, operations, or HR (to name a few). Or HR sees problems from the HR perspective without an understanding of the consequences to accounting, operations, or marketing. For example, poor performance by employees on the assembly line may not automatically be solved with training. From a systems perspective, we might learn that insufficient funding (accounting) has resulted in less maintenance to machines (operations) and fewer changes to product offerings (marketing) as a result of poor top-management support (leading to declining sales).

Looking at the whole versus a part is more holistic in nature. This is an integrated perspective. Even more, it suggests that the goal is not to solve the parts of problems individually within silos, but rather as a whole. Additionally, the collaboration so essential to the DNA of an innovative organization is hindered with a silo mentality.

To be more effective as strategic thinkers, time for reflection is set aside. Whereas the focus is often on doing (versus thinking), strategic thinkers know how critical it is to spend time thinking. And we have already established that innovation certainly requires time for reflection.

Strategic thinkers recognize the importance of asking the right questions. That is, they ask higher-level questions focused on the big picture, which automatically leads them to be more strategic and to address the root causes of problems.

Understanding this big picture acknowledges the importance of the external environment as well. Monitoring trends is paramount for strategic thinkers. Keeping up with current trends can be accomplished by reading broadly (not just our own trade publications) and subscribing to a wider range of e-newsletters and trade publications.

Strategic thinking helps us better address the complexities of our organizations (and our world). Because it requires a long-term perspective, we do not focus just on today and the current crisis. We read the handwriting on the wall and develop peripheral vision while taking off those blinders that constrain our systemwide thinking.

To be more effective in thinking strategically, individuals must be willing to cut their losses and change direction if they are wrong (in light of new information or new opportunities). In other words, they must fail fast (as discussed in Chapter 2).

Strategic thinking, then, is about adding a new lens and generating new insights. In addition, it is about recognizing trends, challenging assumptions, embracing change, and capitalizing on opportunities—all essential to fostering an innovative culture.

Spotlight on Innovation

"Innovation isn't a tick box exercise, or a whim of a tech-savvy HR manager, but a technique to help gain competitive advantage through talent, which remains a company's greatest asset."[11]

Curious Lifelong Learners

An innovative culture requires individuals who are lifelong learners. We need to seek out people who have a natural curiosity for knowledge and a thirst for learning. These lifelong learners enjoy learning for its own sake. They usually possess a wide range of interests.

Naturally curious and inquisitive individuals ask questions continuously. When thinking critically, we must question! Rather than accept things the way they are, these individuals question and challenge. Perhaps most importantly, their problem-solving skills lead them to drill down to the root of problems. There can be no blind acceptance. Questioning gets us to the root causes, helping us frame (or reframe) the problem. But of course, this requires that we ask the right questions. These individuals ask *why not* instead of *why*. This helps to better frame the problem at hand.

Essential to building a culture of innovation is cultivating an environment of questioning. Unfortunately, as adults we begin to pose fewer questions. Adulthood often implies we have more of the answers, and therefore we ask fewer questions and tend to be less inquisitive. Children ask why repeatedly. Organizations must open up to encourage this questioning and the need for lifelong learners. Sometimes we may be better served to think like a beginner.

Even with a depth of subject area expertise, lifelong learners do not think they already know everything. Instead, they recognize how much more they can learn—and are continually open to the pursuit of this new knowledge. They are constantly driven by their natural curiosity.

Lifelong learners, then, are geared toward improvement. They recognize that in a constantly changing world, they must continuously be updating their skills and learning. A failure to do so means that they are falling behind! Their main motivation, however, remains a natural inquisitiveness that must be fed. Confident and successful organizations can and do benefit—in a number of ways—from a workforce of lifelong learners.

Spotting the DNA of an Innovator

A primary element in nurturing our innovative culture is to constantly feed it. Part of this is infusing it with the right "food," that is, hiring people who fit the culture and will continue to reinforce it—but also challenge it. The GIGO concept (that is, "garbage in, garbage out") is relevant here. To derive the output we want from the organization, we must consider the input (specifically, our human capital). Andrew McIlvaine suggested that to be effective, employers must "look for a certain DNA in all our candidates."[12]

We must, then, identify what this DNA is to determine those additional characteristics that we seek in our candidates. In addition to strategic thinkers and lifelong learners, we may desire candidates who are self-starters and who are committed, tolerant of ambiguity, open to change, and adaptable.

Spotlight on Innovation

"The people who are hired and the training and cultural imperatives placed on the business are done so through the role of HR, so HR leaders can have a big impact on whether or not the organization is culturally attuned to innovation."[13]

Committed

Committed individuals are a necessity to our innovative cultures. These are the candidates who demonstrate tenacity and persistence. Only those who are tenacious can come back from failure to try again. They do not give up easily, even when faced with failure.

Creative individuals are committed and resilient. A culture of innovation depends on continuous experimentation, requiring individuals who are committed and live by the mantra "if at first you don't succeed, try, try, again." The acceptance of failure is also essential to commitment. There should be no shame in failing (especially with intelligent failures). The key is to identify those individuals who will remain committed to the quest even when faced with failure and who will bounce back from failure, will learn from it, and will move forward. Resilient individuals are not stonewalled, deterred, or sidelined by failure. Rather they are committed enough to continue the search.

Recent research has explored the significance of "grit" and the role of persistence in staying the course to achieve success.[14] This goes beyond a simple tolerance for failure.

Self-Starter

A culture that empowers its employees must have self-starters to run with that empowerment. Self-starters have the motivation and drive to seek out new ways of doing things. These individuals do not wait to be told what to do, nor do they necessarily want to be told what to do or how to do it.

A flexible structure with less control and autonomy (as discussed in Chapter 2) provides a ripe environment for self-starters. They simply need to be provided general goals and parameters and then allowed the freedom to work within those. Self-starters are ambitious with a strong desire to be productive. Their goal orientation gives them a target for their energy and efforts. And autonomy provides the fuel to their engines—further motivating them.

Tolerance for Ambiguity and Openness to Change

Creative individuals move out of their comfort zones and bravely into the zone of uncertainty. Yet they do not jump to conclusions in the face of uncertainty. And they are not paralyzed by a lack of certainty when there is ambiguity. Discovery Health took that leap into the unknown by reversing the health care model to reward healthy lifestyle changes of its customers. After operating for a little over two decades, Discovery Health is the largest health care company in South Africa. While there may be less certainty and perhaps more risk, the payoffs are huge. Being adaptable and flexible enables creative individuals to see problems from multiple perspectives thus mitigating possible risks.

Spotlight on Innovation

"Fundamentally, innovation is about learning. Inquiry, curiosity, seeking to gain fresh understanding of the problem at hand—these are all vital ingredients for innovation."[15]

Fostering an innovative culture requires attracting people who are comfortable with ambiguity. These individuals want to help shape the future of the organization. They are comfortable not knowing everything with certainty—and even thrive in this environment.

If we are brutally honest, we may admit that some innovative ideas may sound dumb at first. It is this comfort with ambiguity that enables the innovators to take risks. Innovators will even often take the risk of sharing these unusual ideas and risking their reputations because they generally do not have large egos. They do not feel the need to play it safe. Even more, they are okay with appearing "odd" given that they do not worry as much about what others think or say. And of course, in a truly innovative culture, they also do not have to worry if they will be punished in an environment where intelligent failure is embraced.

Professor Brian Griffith and Ethan Dunham, co-founders of Human Capital Performance Partners, described creative people as having a "discovery orientation." They further explained that creative types "do not rely upon proven methods or established procedures" and are "not bound by convention."[16] Some researchers even depict creative types as being a little quirky.

Not only are they open to change, but they take it further to embrace change and even create this change. As a result, they are willing to experiment and are not satisfied with the status quo. Their openness to change also makes them comfortable rocking the boat— voicing concern and tactfully creating conflict around ideas. Being open to change also enables people to be more open to feedback.

Adaptive individuals consider (and are ready for) unintended consequences. There is a courageous recognition that they cannot think of everything. Furthermore, they are also open to being wrong. This requires a level of self-awareness; that is, they are aware of their own biases that may create blind spots. As a result, they are comfortable challenging their biases and assumptions.

Relationship Builders/Collaborators

Creativity does not usually happen in isolation. The collaborative element of an innovative culture is based on relationships and support networks. Social contexts, then, bolster the creative process.

This is how ideas get nurtured. Innovators thrive on the opportunity to connect with like-minded people.

Employers must seek out individuals who are comfortable interacting with others outside their own functional areas, organizations, or even industries. Individuals who can collaborate across lines and build relationships across disciplines are essential to the innovative culture. That creative individuals are often great collaborators benefits organizations because working in team environments requires well-developed collaboration skills. Using networks to exchange ideas and brainstorm stimulates the creative process. In addition, relationships help in garnering support once those ideas are generated and are ready to be implemented.

Those with a tendency to hoard information are not a good fit for a culture that requires collaboration and an open sharing of information. These individuals may not be the best team players.

Spotlight on Innovation

"Creativity is intelligence having fun."[17]

Bringing It All Together

Although we have discussed the skill set and characteristics needed to help guide our innovative culture, we do not necessarily want to think in terms of a checklist of characteristics to use when recruiting innovative individuals. Instead, we want to seek out individuals who possess a complete portfolio of skills that enables them to be innovative. That is, we want to focus on the whole individual—or the package—while searching for these key characteristics. We may seek out these characteristics individually, but it is how creative people use these characteristics together that is important. Bringing these characteristics together requires sound communication skills and a positive attitude. These two may be thought of as the levers by which the other characteristics are executed and come together to deliver results. That is, they may be the oil that greases the innovative engine as all the characteristics (the gears) are integrated.

Good Communication Skills

Effective communication skills enable individuals to present their ideas, persuade others to gain buy-in, and deal with conflict. In addition, strong communicators are able to share knowledge and information with others. Their refined communication skills allow them to challenge ideas and even rock the boat to confront others. They gain buy-in and support for their ideas by competently engaging in persuasive communication. Without strong communication skills, the best ideas may go unnoticed and never see the light of day.

Positive Attitude

Optimistic thinking is a critical innovative thinking tool. Creative individuals possess a passion for what they do. This optimism is contagious and often inspires others. Their positive attitude communicates a sense of "can do" that encourages others to support their ideas. We want to recruit those who engage in optimistic thinking. These are the individuals who are less likely to quit when faced with failure or adversity.

Furthermore, creative individuals have fun! People are more likely to be "in the zone" or experience "flow" with high levels of engagement when they are passionate about what they are doing. And this energy is contagious! Research has even confirmed that a sense of humor is positively correlated with creativity. Bottom line, a healthier positive environment is created with optimistic individuals.

The Interview

Since we have established that our innovative culture is a marketing tool, we need to communicate that. We might consider demonstrating some of this innovation in the recruitment process itself. William Sebra highlighted the importance of the "need to do more to demonstrate innovation in recruitment and talent management."[18] To retain our talent, then, nearly half of the approximately 5,000 global respondents in the Innovation Imperative Study reported that they "valued direct contact . . . through email or services such as Skype, as well as the use of professional networking tools such as LinkedIn."[19] We need to walk the innovative talk in all that we do in HR to market our innovative culture. (There is more in Chapter 7 on innovative HR practices.)

We might consider asking applicants to share something extraordinary that they have done to gain an idea of their comfort zones and their motivations. We might ask candidates for examples of their entrepreneurial spirit. And we can ask about their favorite books (being sure to follow up with asking why). By the way, if you ask this be prepared to discuss *your* favorite books as well. We could even play with analogies to see how they think.

Though HR professionals often conduct the majority of the formal interviews, an innovative culture is fostered when everyone in the organization is continuously seeking out new talent. Referrals can be a great source of recruits—when embraced by all.

But we must be careful what we promise. The realistic job preview (RJP) is highly recommended. That is, we want to paint a realistic picture of our culture—as it is and not as we hope it will become.

Chapter 4

The Awakening: Developing a Workforce that Embraces Innovation

"Organizations cannot expect to be innovative and successful unless stakeholders at each level acquire the knowledge and learn the practical skills and behaviors necessary to make their organizations more innovative," according to the ASTD white paper *Building an Innovative Organization: The Role of Training and Development.*[1] HR's responsibility is to integrate those elements of an innovative culture into the training and development function across the organization. Scott Edinger made a case for the long reach of training and development's impact in saying that "a culture where innovation thrives in every corner is exponentially more valuable than a culture which anoints one or even a few people as the innovative ones."[2]

To develop a workforce that embraces innovation, it is essential that the organization understand the process of innovation itself and the characteristics of the innovative individual. A commitment to continuous learning and a focus on creating lifelong learners is required. The foundation for developing an innovative workforce is building on the natural curiosity of people.

The People Factor

According to Robert Todd and Lisa Buckley, directors at ?What If!, and Sam Herring, co-founder and CEO of Intrepid Learning, "If innovation is not already part of an organization's DNA, that capability must be developed."[3] There can be no denying that people are a primary focus of the innovative culture. It is impossible to be innovative without people—and people who are appropriately developed. The innovative culture works in tandem with the development of an innovative workforce. A culture closed to innovation will not allow most innovative individuals to thrive. And the organization must provide the skills for the individuals to develop and grow.

Spotlight on Innovation

"Managers can in fact nurture and promote creativity in employees who are not naturally predisposed to be creative."[4]

It is an iterative process. As we recruit innovative individuals, we must commit to their continuous development—providing the tools they need to be innovative. Although we certainly can (and should) hire innovative individuals to foster our cultures, we must also provide these individuals with ongoing developmental opportunities after being hired to "feed" and nurture them. And equally important, we must develop the current workforce so employees may be equal collaborators and contributors in the innovative process.

In developing a workforce that embraces innovation, organizations must balance a short-term focus with a long-term focus. We can easily get caught up in the short term, but a long-term focus is necessary if innovation is expected to thrive. Unfortunately, the tendency is for businesses to squeeze their margins and focus on the short term. Creating a leaner business, however, sets us up to lose critical talent and allow other talent to become obsolete and stagnant. These are not the individuals driving innovation.

According to the authors of the article "Drive an Innovative Culture," "about 84 percent of the variance of innovation effectiveness could be explained by organizational culture. ... Results pointed to the importance of adaptability and mission as strong cultural innovation drivers. ... Organizations can enhance innovation by empowering and developing their people."[5]

To truly produce an innovative culture, employers must provide individuals the appropriate development. This is an ongoing, long-term investment in our organizations' futures—one which requires the commitment of top management. Innovation is recognized as a people issue, then, and is closely aligned with the learning function. Our responsibility is to teach everyone, especially our front-line employees, how to innovate.

A willingness to innovate is the first step in developing our workforces. Just as we can lead a horse to water and not make him drink, so it is with developing an innovative workforce. Organizations must help employees understand why being innovative is their job—and how it can benefit them. The organization's leadership must drive this communication as it continuously addresses innovation and the business case for it.

Spotlight on Innovation

"There is now general recognition that the innovative potential of an organization resides in the knowledge, skills, and abilities of its employees."[6]

Sharing the Big Picture

As the need for innovation permeates the organization, we must truly make it everyone's job, from the CEO and top management to the front-line employees. Doing so requires that everyone knows where the organization is going. That is, everyone must understand the big picture and, most importantly, how they fit into that big picture. Connecting the dots for people to understand how what they do contributes to the overall corporate objectives provides the direction so essential for everyone.

Understanding the mission and goals of the organization ensures that the workforce is unified in that one direction. Though the stated goal is not innovation, the result of attempting to meet our goal may result in innovation. Innovation for its own sake is not of value to us. Instead, it is valuable when we are able to perform more effectively in fulfilling our mission. What drives our innovation, then, is a clear understanding of our mission. We can become more innovative as we better understand where we are going and what is expected of us, that is, the big picture.

Providing individuals with the opportunity to rotate through a variety of positions or functions within the organization allows for the ability to better see and understand the big picture. (It also helps us ensure that people are matched with the most appropriate positions to best use their talents.) Without an understanding of the interconnectedness of the parts of the organization, we suffer from tunnel vision and a silo mentality. Learning about other areas of the organization can also be energizing as we recognize the importance of making connections between seemingly unrelated parts.

Encouraging a New Kind of Thinking

A barrier to innovation is thinking that we are not creative thinkers. Developing creative thinking skills in individuals goes a long way toward helping individuals see that they can become more creative!

Perhaps most difficult is the need for us all to embrace paradoxical thinking. The irony for everyone today is that solving problems is not always about seeking out the one right answer. Business thinking traditionally has been more logical and based on precedent. This creates an inherent paradox for us. Our businesses want decisions made swiftly, often in response to the crises of the day. These situations are becoming increasingly complex with no clear path forward. That is, there is no clearly identified one right answer. Innovation requires that we slow down to think more unconventionally and avoid making decisions based on precedent—or the way we have always done things.

Traditional business thinking uses "either/or" thinking to arrive at that one right answer. Innovative thinking is almost the polar opposite of the traditional thinking in business, recognizing that there may be more than one right answer. Innovative thinking resides more on the right side of the brain—using more intuition—rather than the rational, logical processes found on the left side of the brain. Innovative thinking embraces the ambiguity of a new situation, whereas traditional business thinking seeks out precedent and finds comfort in "the way it's always been done." Innovative thinking seeks out new and better approaches by asking "what if?" questions that march optimistically into the unknown. It is our assumption about how things are supposed to work that often holds us back. We need to take a strategic view and rethink what we think we know.

Spotlight on Innovation

"Deconstructed work is revolutionizing talent management, too, and it requires leaders to approach decisions with advanced tools and a keen sense of how pieces fit together."[7]

In a driving analogy, Gary Hamel and C. K. Prahalad (in their book, *Competing for the Future*), suggested that organizations may need to consider avoiding pressing harder on the accelerator and instead consider switching gears. That is, we have an inherent tendency when a strategy is not working to simply respond by doing the same thing we have always done—just bigger, better, and faster (pressing harder on the accelerator). Instead, Hamel and Prahalad recommended that we consider trying something different (shifting gears).[8]

As the challenges of the dynamic business landscape become increasingly complex, it makes sense that traditional business thinking will not suffice. Though traditional thinking may be appropriate at times, new challenges often require new approaches—or innovation. Unconventional thinking is needed. Ironically enough, it is the very absence of the clear path forward that enables us to open the door to innovative thinking by posing the critical question "what if?" One relevant and insightful example recounts that "In the early 1990s Microsoft was developing a technology to compete *with* the Internet. One day, Bill Gates reversed the effort, brought the Microsoft leaders and software developers into a room, and directed that they drop the project and start working on projects to *build* the Internet."[9]

Development in the power of observation is essential to innovative thinking. To be a creative thinker requires that we engage in observation and reflection. The key is learning to see what others do not see, especially details and patterns. This requires taking the time to slow down and observe from different perspectives. We should listen carefully and with an open mind to others to understand their perspectives. In addition, we must learn to go with our gut—that is, trust our instincts and intuition. This is all about the struggle between the left and right sides of the brain.

To become better observers, Daniel H. Pink (author of *A Whole New Mind*) proposed that we sign up for art classes.[10] Drawing helps us use different parts of the brain (and those primarily on the right side). Although we do not necessarily need to offer art classes for our workforces, we *do* want to consider providing developmental opportunities focused on embracing paradoxical thinking and observation skills.

Challenging our thinking is crucial. This may include the need to challenge the taken-for-granted assumptions about what we do and how we do it. Learning to reverse our thinking can pave the way for innovation. Tina Su discussed an idea from author and speaker Scott Berkun to focus on the solution to the opposite of the "real" problem.[11] This enables problem-solvers to flip their thinking. This approach can open the door to humor and more creativity when working in a team. For example, the team is trying to design a better smartphone, reverse thinking would have the team brainstorm ways to design the worst smartphone.

Innovative thinking can give us permission (and even encourages us) to have fun. It is this playfulness and freedom from seriousness that allows us to test odd ideas and ask those wild, "what if" questions. IDEO has embraced the use of playfulness in the innovative process—even poking fun at its own organization.

Spotlight on Innovation

"When you integrate this intuitive ability with learned information and knowledge, you operate using all your resources which provides flashes of insight and recharges your thinking."[12]

Learned Optimism: A Skill to Teach

The good news for our organizations is that whereas some individuals are simply born as optimists, individuals can learn to be optimistic. We can provide advice to our workforces to develop an optimistic attitude—or learned optimism.

Bruna Martinuzzi, founder of Clarion Enterprises, suggested that "optimism is an emotional competence that can help boost productivity, enhance employee morale, overcome conflict and have a positive impact on the bottom line."[13] Described as a powerful tool for leaders, optimism is valuable in encouraging people to move away from the conventional ways of working to attempting new processes.

Alan Loy McGinnis identified 12 characteristics possessed by optimists. Two of these are particularly interesting when understanding their roles in the innovative process. Of most significance is the optimist's tendency to work incrementally toward total solutions by seeking out partial solutions. According to McGinnis, optimists also are always stretching for that ever increasing benchmark of their personal best. In addition, optimism enables individuals to persevere and bounce back from failure.[14]

Motivational experts have recommended for decades that we consider our environment when trying to address our attitude. To be more optimistic, we want to be sure individuals place themselves in "target-rich" environments. That is, surrounding ourselves with more optimistic individuals provides a greater likelihood that we will also be more optimistic. Because attitude is contagious, spending time with pessimists on a regular basis can hinder our ability to be more optimistic. Surrounding ourselves with like-minded, creative individuals also provides a forum to talk things through when we are stuck. They can provide a sounding board.

Part of being optimistic is being aware of our language. Positive language can reinforce optimism and provide encouragement to others. Negative language in a public forum at work can create a more pessimistic environment that is not open to ideas and that sucks the life right out of a room and the individuals in it. Our own internal negative dialogue ultimately influences our behaviors and actions. Thinking negative thoughts is likely to manifest itself in negative and pessimistic actions, bringing down those around us. The power of positive visualization is familiar to star athletes as they are told to visualize that "Super Bowl moment" or crossing the finish line first. But this approach does not apply only to athletes—we can all visualize that "winning moment."

Spotlight on Innovation

"A pessimist sees the difficulty in every opportunity; an optimist sees the opportunity in every difficulty."[15]

Teaching people to be more mindful of their language and to use more positive language involves shifting from "yes, but" to "yes, and." One small word can make a big difference. "Yes, but" is more negative in nature and opens the door for defensive communication. Listeners often hear only what comes after the "but," which generally refutes what they initially said. "Yes, and" is more agreeable and sounds like we are on the same page as collaborators.

In addition, self-limiting beliefs must be tackled. The number one hurdle to creativity is when people think they are not creative. If we believe that we are not creative, we most likely will not be creative. In *7 Habits of Highly Innovative People*, Tina Su posited that "innovation is more about psychology than intellect."[16] The Pygmalion effect (also known as the self-fulfilling prophecy) is at play. Optimistic expectations often yield positive, optimistic outcomes! And the reverse is also true: Pessimistic expectations often yield negative, pessimistic outcomes.

Marcus Buckingham advised us to play to our strengths to enhance our optimism.[17] Focusing too much on the improvement of our weaknesses can bring us down. This occurs as we become physically and mentally drained performing those activities that do not allow us to use our strengths. Buckingham described these as the activities that deplete us. As HR professionals fostering an innovative culture, we want to help individuals find those activities that help them be "in the zone" and that have the opportunity to play to their strengths.

Tools to Consider

Some skills and abilities essential to innovation are more innate. The good news is that other skills can be developed in our workforces.

Foremost, when developing an innovative workforce, we must be sure employees are ready to be empowered. Providing freedom in the innovative culture is critical (as discussed in Chapter 1). To execute on this empowerment, then, individuals must first be competent. The foundation for this is specific expertise.

People are naturally more likely to be innovative when they are motivated by their work. A passion and natural curiosity often provide the motivation for our innovative efforts. Such was the case with Alexander Graham Bell. His motivation to explore acoustics was provided by his mother's increasing deafness. And then his continued experimentation (and failures) led ultimately to the telephone. The lesson for us here is to align people with the jobs that engage them—and to keep them engaged. Career development plans and career paths should incorporate these issues.

Spotlight on Innovation

"Innovation may be today's most important business competency."[18]

We can all benefit from identifying where our "creative zone" is. Some recognize that when they are sleeping they are creating and should therefore keep a pencil and paper near the bed to capture those ideas. Others realize their creative insights occur in the shower, or when driving to work, or when jogging. Wherever we experience our "eureka" or "aha" moments, we need to learn to pay attention to them and capture them. That is, we need to leverage that trigger to our innovative ideas. Some people keep journals, others keep folders with scraps of paper dropped in, and others use recorders or smartphone reminders. Whatever works for each is best. The key is to capture these ideas and then continue to review them.

Suggesting that people keep an idea log or file can be a great strategy. It is particularly helpful if people come back to regularly revisit their ideas. Jack Dorsey's idea for Twitter was initially developed in 2000. He waited eight years for the technology to be developed to execute the refined idea of a broader audience for instant messaging.

The complex challenges our organizations face today can be daunting and overwhelming. Innovation often thrives if we take small steps and break the bigger issue into smaller pieces. Closely related to this concept of breaking our bigger challenges into manageable pieces is to be reminded to heed the small problems and to develop solutions for them before they become those daunting, bigger challenges. Deconstructing a larger problem into small pieces may provide us with a new perspective—and an innovative solution not readily apparent. This deconstruction requires an understanding of all the parts and of how they fit together.

Given that innovation often focuses on solving problems, development in problem formulation skills is helpful. Changing the way a problem is stated can often alter our approach to a solution. Perhaps even more critical, problem identification skills should be addressed. Whereas rational decision-making has been emphasized in the past, developing skills in decision-making might focus on scenarios. The use of best-case, worst-case, and most likely outcomes can be helpful.

As difficult as it is for many of us, we must fight the urge to censor our ideas. Too often we fall into the trap of considering what is right or wrong, and then we lose the best ideas. Workshops using brainstorming techniques can help us manage this tendency and aid us in generating a larger number of ideas.

Providing development around some general competencies is also worthwhile. Effective innovative thinkers also practice collaborative inquiry. Because innovation is usually a team sport, it is necessary to open a dialogue with others to share ideas. Development in communication skills can be beneficial. Presentation skills can be helpful in enabling individuals to better pitch their ideas to gain buy-in and support from others. Due to the collaborative nature of the innovation process, development in team dynamics is valuable. Topics might include emotional intelligence, communication skills, conflict resolution, and networking. Furthermore, we must practice active listening, which enables us to reflect back what we have heard and build on the ideas of others. Teaching individuals self-reflection also includes nurturing the spirit with doses of inspirational readings.

Spotlight on Innovation

"Even in the most constrained environments, innovation capability can be grown and can yield results."[19]

In developing an innovative workforce we may want to consider offering developmental opportunities for individuals in resilience, pattern recognition, experimentation, brainstorming skills, and coaching in the creative process. Mind mapping is a tool that might be explored.[20] It enables a visual depiction of our ideas that can be used in brainstorming sessions.

Creativity/Innovation as a Leadership Competency

Critical to the development of the workforce is the development of leadership. Because we must match organizational needs with top management's talents, we must consider leadership development with an emphasis on creativity and innovation as valued leadership competencies. This sends a clear message that the organization values innovation.

The leadership talent pipeline must be filled with creative types. Just as a successful sports team must have bench strength, so it is with our organizations. And when it comes to identifying our leadership high potentials, we want to be sure that we are using the competencies required for success in an innovative culture as our guides. Traditional competencies may no longer be as effective when identifying our leadership talent.

To foster an innovative culture, we must address the skill set of leadership to lead the way. If the organization's leadership is not innovative and skilled in managing innovators, the potential for advancing that culture is lost or at least is greatly diminished. Leaders in an innovative culture must be comfortable developing others in critical competencies, working in a more flexible environment, providing autonomy to others, and managing innovative teams. Leaders at all organizational levels should be prepared to lead teams across disciplines and functions. As mentors, they should be prepared to provide support networks and the opportunity for intellectual stimulation. Perhaps most importantly, they are responsible for providing the opportunity for collaboration and the environment whereby innovators can become self-motivated.

There is more on managing others in Chapter 6.

Spotlight on Innovation

"Companies that cultivate leadership skills are more likely to net successful innovations."[21]

Chapter 5.

The Carrot and the Stick: Rewarding for Innovation

The recognition and reward system plays a major role in the ability of the organization to move more toward an innovative culture. We must be innovative in our approach to rewards and acknowledge that there is no one-size-fits-all option. The implications of intrinsic and extrinsic rewards are discussed with an examination of financial and nonfinancial options.

Because we get what we reward, care must be taken in managing the reward and recognition process. "The goal in creating a culture of innovation is to be able to link the generating of good ideas to the bottom line."[1] It is essential to clearly communicate to the workforce the criteria by which ideas will be evaluated. The approaches used by some firms today are discussed.

Sustaining Innovation

In effectively managing any organizational culture, a key element to explore is how rewards are allocated. We established in Chapter 2 that a critical function of culture is to shape behavior and reinforce our values. As a result, the development of our reward and recognition programs takes on added importance. We must ensure that we are rewarding the behaviors that we say we value—and that we are doing so in such a way that is consistent with our cultures.

Spotlight on Innovation

"Without the hope that something might actually be done with their ideas, there is little motivation for people to participate."[2]

At the core of sustaining innovation, then, is addressing the reward and recognition of our workforces. Reinforcement theory reminds us that positive rewards following behaviors increase the probability that those behaviors will be repeated. So if employees engage in innovative behaviors that our organization values and they receive positive rewards, they are likely to continue to perform those innovative behaviors. In addition, others in the organization who see employees being rewarded for these behaviors are likely to learn that innovation is valued. And they then may also begin practicing these innovative behaviors to be similarly rewarded and recognized. According to Ann Bares, managing partner of Altura Consulting Group, "High performers are more than twice as likely as low performers to report that their organization rewards innovation."[3] Crafting an appropriate reward or recognition program enables us to further engage and retain our innovators.

One organization, the American Productivity and Quality Center (APQC), "has found that to drive innovation in products and services, an organization needs innovative approaches to rewards and recognition."[4] We must think creatively in the development of these programs and be innovative in our approaches. We must appropriately design reward and recognition programs that motivate individuals to produce the outcomes that we seek.

Ironically enough, although most companies know they should be addressing an improved approach to formally rewarding innovation in their organizations, a whopping 90 percent do not actually do so![5] A fundamental foundation for an effective reward program in an innovative culture recognizes that good ideas provide value— and should be recognized. The objective of an effective reward and recognition program is to build our pipeline of innovative ideas.

Reinforcing the Culture

The culture literature reminds us of the important role of the visible aspects of culture that reinforce our values. These include artifacts or the symbols of our cultures. Certificates, medals, and even photographs become proud symbols emphasizing our values. As these are prominently displayed for organizational members to see, our values are reinforced. The Procter & Gamble awards for failure serve to reinforce the permission to fail in the organizational culture. The Nike Winnebago (discussed more in Chapter 7) serves as an artifact communicating the value placed on innovation. Recognition of iconic figures such as 3M inventor Arthur Fry subtly affirms our organizations' cultural values.

The role of stories is also relevant in reinforcing company culture. Sharing success stories is a meaningful component in communicating what we value. As these stories are told and retold, there is little need to tell new employees that innovation is valued. The story does it for us! Showcasing our innovators can highlight them as role models to illustrate the value that we place on innovation. Fry and the Post-it Note story at 3M has become a classic in communicating the value that the organization places on innovation and on its ability to embrace failure (and celebrate it). The communication of this story even goes beyond 3M's organizational boundaries.

Spotlight on Innovation

"People will be most creative when they feel motivated primarily by the interests, satisfaction, and challenge of the work itself—and not by external pressures."[6]

Extrinsic versus Intrinsic Rewards

When developing a reward and recognition program, we want to consider the interaction of intrinsic and extrinsic rewards in motivating individuals. Extrinsic motivation occurs when individuals engage in behaviors to gain an external reward (or in some cases to avoid an external punishment). Intrinsic motivation occurs when individuals engage in activities for their own personal fulfillment or enjoyment.

Research has revealed that extrinsic rewards must carefully address this interaction when providing rewards.[7] We may want to exercise some caution when using rewards. The extrinsic rewards may actually undermine intrinsic motivation and may ultimately become less motivating over time. This, then, highlights the importance of hiring people who are intrinsically motivated (or otherwise self-motivated) by the work (as discussed in Chapter 3).

It is possible, however, for the organization to enhance an individual's intrinsic motivation. Providing recognition in the form of praise can heighten intrinsic motivation—through the enhancement of self-confidence and the perception of competence. The culture itself may provide some of that motivation or at least sustain the environment where the individual can be self-motivated, that is, where the intrinsic motivation is allowed to happen. Knowledge fairs for collaboration held by the World Bank and 3M's famous "15 percent rule" (which provides time for innovation) are great examples of workplace environments that provide a forum for intrinsic motivation. Cultural elements such as empowerment and freedom may provide intrinsic motivation to individuals.

Placing people in positions that are aligned with their skills and strengths can be one of the most powerful intrinsic motivators. Using stretch goals and then giving employees the autonomy to determine the way in which the work will be done increases motivation. And providing the organizational support (in terms of the resources needed) and managerial support can further motivate our innovators.

According to Alex Gammelgard, director of product marketing at Apttus, "Humans are born with the innate desire to problem-solve, which can translate into innovation."[8] The canvas is there. We just need to take advantage of the opportunity to write on that canvas, thereby developing that translation. And we do so by fostering an innovative culture with an appropriate reward and recognition program that communicates how much we value and appreciate those who innovate.

Intrinsic motivation comes from within. Individuals are more likely to innovate when they see a problem, understand it, and are empowered to own it. People get better at problem-solving the more they practice it. As they improve at finding solutions with smaller problems, individuals tackle larger and larger problems. They can build on these mini successes to develop more self-confidence in themselves and in their problem-solving abilities. And of course, through the process their successes often provide more motivation to continue applying innovative techniques to their problem solving. We have a tendency to participate in activities in which we experience success and are more likely to experience higher levels of job satisfaction. Creating environments where individuals can be self-motivated can enhance our ability to innovate.

Spotlight on Innovation

"Several innovative organizations have encouraged peer recognition, arranged events, and established work structures conductive to cultivating relevant innovations."[9]

No One-Size-Fits-All Programs

It is critical to fit our reward and recognition programs to our organizations, our people, and our cultures. As a result, there is no one-size-fits-all approach. A lack of alignment results in an ineffective program at best with few of the desired results achieved.

A paradox of rewarding exists in that we want to be consistent in rewarding individuals, and yet one type of reward does not work across all our departments and businesses or for all individuals. A team is composed of individuals who do not all respond in the same way to one specific incentive. Employers must know people as individuals and learn what they value to provide the most meaningful rewards and recognition.

Providing employees with the opportunity to attend innovative events even outside their own organizations encourages them to engage in similar activities in their own jobs. Making the pursuit of innovation fun encourages more individuals to experiment. Box.net rewards ideas with pizza and beer during an all-night brainstorm/ hack fest. Google provides one day each week for individuals to pursue their own projects—not necessarily those included in their regular job responsibilities.

Therefore, in crafting a program that really works to achieve our desired outcomes, we need to ask for input from employees themselves. Investing in a program that employees do not value is worthless. Providing employees with choices is most likely to be effective in meeting our objectives.

Additional opportunities may be provided as rewards for innovators. Entrance into management development programs provides a great reward—and is often highly visible. An opportunity to facilitate innovation courses provides individuals the chance to share ideas and approaches with a broader audience, and again, to increase visibility.

If we truly embrace failure as part of our innovative culture, we may want to consider *how* it is incorporated into our reward and recognition programs. A safe environment that does not punish failure, but rather encourages individuals to learn from it, may choose to reward failure! We want to avoid punishing individuals for trying and failing—that is, for doing something. Any punishment (if necessary) should instead target people for doing nothing.

The key to the most effective reward programs begins with truly understanding the motivation of the workforce and identifying our objective. That is, what is it we hope to achieve as a result of our program? The focus should be on filling the pipeline with ideas. To be most effective, a reward system must be carefully designed to align with our objectives and values. Decisions must be made about who will make the reward decision, what ideas will be included, whether only results (not effort) will be rewarded, and how we will use financial and nonfinancial rewards.

Spotlight on Innovation

"55 percent of employees strongly agree that the quality of their company's recognition programs affects their performance, but only 10 percent of those polled are satisfied with these efforts."[10]

Financial versus Nonfinancial Rewards

When deciding to offer rewards, we must then make decisions about the use of financial and nonfinancial rewards. It has been reported that 57 percent of organizations identified as high performing use nonfinancial rewards when recognizing innovation.[11] Research continues to confirm that money is not the number one motivator! So it would make sense to build in at least some nonfinancial rewards.

3M opted out of using monetary rewards because over time they can undermine the intrinsic motivation to innovate. Peer recognition has become the reward of choice for 3M. Peers select the recipient of the Technical Circle of Excellence award. 3M also provides grants to employees if not funded through "normal" channels. Covert recognition may involve simply moving a project idea to the next phase of development.

And yet, bonuses remain the most common reward offered across most organizations. The trap we fall into with money, however, is that the amount of the reward may not always be aligned with the actual value of the idea or innovation project. We also run the risk when using monetary rewards of simply running out of funds. When this happens, does innovation grind to a halt? Research on continuous reinforcement schedules would say the answer is yes! So we may want to consider a balance that includes nonfinancial rewards.

Some of the tangible rewards offered may include money, gifts, certificates, plaques, iPads, and gift cards to restaurants or local malls. Some may even provide premium parking spaces for designated periods of time—often with a sign announcing what it is for. Extra paid days off have become increasingly popular. A three-week vacation is even offered at Firstborn. Some companies allow employees being recognized to spin a wheel for prizes of varying value.

Flexibility and choice can be built into programs. Companies can use a point system to allow individuals to earn points that can be redeemed for "gifts" or services from a catalog. Frima's point reward program is used to reinforce its culture of work/family balance. The company offers "family" services such as babysitting and home repairs to free up employees' time.

Nontangible rewards may include praise, public credit, parties, public ceremonies, and dinners recognizing award recipients or perhaps publicity in company newsletters or special announcements on the company intranet. Announcements on company billboards or electronic signs may be used. Other ideas include certificates of appreciation or a dinner or lunch with company leadership, professional development opportunities (such as conference attendance), or a company "hall of fame" induction. Lapel pins have been used effectively for decades as a special recognition tool by fraternities to communicate elite membership. Companies have begun to adopt this as a way of building a "community" membership. As another type of reward, some companies provide employees with the opportunity to select the next work assignment for themselves.

Spotlight on Innovation

"Non-monetary awards such as *recognition in the organization* and the *actual realization of ideas* are considered as more important."[12]

Who and When to Reward/Recognize

Members of the workforce need to understand the criteria on which recognition and rewards will be decided if they are to place trust in the system. Companies, then, must determine who makes the decision of who gets recognized and when the rewards will be given. This includes the decision of how often ideas will be evaluated and when (and how often) recipients will be acknowledged.

The question of who makes the decision regarding who will be rewarded and recognized is not as straightforward as some might like to think. We have a number of options to consider. The recommendation can be made by managers, peers, or employees themselves (through self-nominations). We may naturally assume that managers make these decisions. However, programs whereby individuals are nominated by their peers can be quite effective. Peer recognition is extremely important. Ironically enough, in larger organizations where the tendency has been to provide recognition from top-level executives, research has pointed out that people value recognition from those closer to them, especially their direct supervisors.

Team rewards should celebrate the entire team and its efforts—not just a select few individuals. Because innovation tends to be more of a team activity, it is difficult in some cases to identify who should be rewarded. The innovative process may involve those who develop the idea, tweak the idea, execute the idea, and provide routine maintenance of the idea. Who to hand the check to or whose name to include on a certificate is not always an easy answer and can even create some hurt feelings and conflict. It is imperative that we provide rewards to employees who work together, especially across departments. We must beware of rewarding individuals when innovation is the work of a team. We can miss valuable opportunities to acknowledge and encourage teamwork. Two important by-products of team recognition are the reinforcement in the value of membership on that team and the increase in team cohesiveness, both of which can boost team performance.

Companies can choose to reward at multiple times: when the idea is conceived, when it is approved, or when it is implemented. As the reward and recognition program grows, companies tend to acknowledge individuals less often—perhaps annually. When launching a program, it may be kick-started with more frequent acknowledgements. And of course, we can use spot rewards (such as gift cards) with just a small budget to keep the program visible and to continually prime the pump of innovation.

Spotlight on Innovation

"The right rewards system provides a powerful force for reinforcing commitment, directing employee professional growth, and shaping the corporate culture to be more innovative. HR departments must look at the reward mechanisms in place and ask if they are doing the right things to develop the employees and culture of the organization."[13]

How to Reward/Recognize: It's Not Necessary to Go Big!

It is not always necessary (nor even effective) to provide large rewards. Verbal acknowledgements can be very powerful and should not be overlooked. A handwritten note by a member of leadership goes a long way. Simple recognition is one of the most effective motivators.

There is concern (and rightly so) with the effect of overcompensating. Because selecting "winning" ideas for recognition is difficult, some companies opt to reward all innovative ideas. The FruitGuys is one such company. The danger, though, is that we may reward people for what they are already expected to do. Some companies decide, therefore, on specific standards or benchmarks required for ideas to be compensated.

On the other hand, a reward and recognition program can also create a healthy environment of friendly competition among innovators. It can create a sense of membership that people strive to achieve. Contests have been used effectively to generate healthy competition. In addition, an opportunity to showcase innovative ideas can be motivating. Foursquare holds "Demo Days" weekly to allow employees to share their ideas and receive feedback.

Above all, the system must be fair. We do not want the system to be structured in such a way that it becomes divisive. It is deemed unfair if some managers use it more than others—or have more opportunity to use it than others. The fairness of the system is often ensured when we communicate the criteria. It helps to have "alignment throughout the company of what is being done, and that involves everyone at the company at some level."[14] Projects with high uncertainty may possibly be measured or evaluated differently in respect to their progress. This is fine as long as we are sure to openly communicate the process to everyone. Employees are more likely to deem the process fair if everyone understands the criteria up front.

People need to know where to go with their ideas and how they will be evaluated. We may also want to address whether they will obtain feedback and, if so, what the process may look like. It is important that we be clear about our criteria for evaluating ideas. This should include how we will address incremental ideas.

Spotlight on Innovation

"Culture is a critical factor in promoting innovation. Business leaders stimulate innovation by offering incentives to workers, creating an environment, and setting expectations."[15]

A Final Caution: Revisiting the Implications of Reinforcement Theory

In closing this chapter, a final caution is in order.

The old saying "be careful what you wish for" certainly applies to some organizations that have not fully embraced innovation. We know we want to be innovative and often talk a good game. And then, when we receive the innovative ideas, we are not sure what to do with them. We got what we asked for and then are somehow surprised, and even more, we are afraid of these new ideas that move us out of our comfort zones. We are not able to take that last step to execute. So we do nothing—no reward, no recognition, no acknowledgement. Or equally demoralizing, we censure—sometimes harshly—the ideas we receive without providing constructive criticism. This destroys the culture of innovation as it goes against the very foundation. Then we act surprised when people stop engaging in innovative behaviors and our top innovators begin to leave the organization.

An innovative culture needs to manage a pipeline of ideas—and obviously have a system to evaluate them. Given that we get what we reward, a system that rewards and recognizes innovation is more likely to produce innovations. That is, we must develop a system to screen and rate ideas. The reward and recognition system that we create then reinforces our culture and communicates the importance of innovation.

Chapter 6.

Managing People in an Innovative Culture

Key to the management of an innovative culture is an understanding that innovation comes from everyone, everywhere in the organization. Human capital is managed differently in the most innovative organizations. Encouraging innovation in our workforces requires different approaches to people.

An innovative culture is best served with an emphasis on outcomes and a different focus on goal setting whereby stretch goals are emphasized. Topics discussed in this chapter include creating a safe environment to fail, providing transparent communication, offering opportunities for collaboration, allowing time for reflection, and reserving a "space" for innovation.

Empowered employees are critical to the innovative process. This empowerment only happens with top management's commitment to model giving up control and investing in the development of the entire workforce—and creating fluid structures where innovation can thrive.

Don't Forget Management Innovation

Not only must we be aware of how to manage our workforces to foster an innovative culture, but we must also consider management as an area for innovation. That is, we must be innovative in our approach to management as well. Management thought leader Gary Hamel reminded us that we have a tendency to explore more processes for product innovation without considering management innovation.[1] Forgetting management innovation is a missed opportunity.

Toyota provides a good example of a company that leveraged its management innovation to become more competitive and efficient. Moving away from the traditional model of using staff specialists in process improvement, Toyota provided the necessary skills and resources to all their employees to be proactive to identify problems in advance and fix them when they did occur. Toyota specifically empowered their employees and provided them the resources to act.

Spotlight on Innovation

"Savvy leaders shape the culture of their company to drive innovation. They know that it's culture—the values, norms, unconscious messages, and subtle behaviors of leaders and employees—that often limits performance."[2]

And the list goes on . . . Whole Foods challenged traditional management approaches by using an empowered team to manage each department in its grocery stores. Whirlpool provides yet another good lesson for those building an innovative culture. Since the goal is "innovation from everyone, everywhere," we need to build the value of innovation into all the organization does. This includes the leadership development program. Whirlpool's program included innovation mentors to reinforce the value placed on innovation.

Leadership can provide a positive role model by being adaptable—even in its approach to management itself. A self-assessment can be useful to identify the management processes that shackle creativity. (The barriers to creativity are discussed more in Chapter 8.) As leaders in innovative cultures, we must model the behaviors that we say we value and that strengthen the culture. This requires that we continuously communicate the importance of innovation in all we do and be active in creating the infrastructure that supports innovation.

Creating a Safe Environment to Fail

An essential element of an innovative culture is an environment that allows people to take risks and to fail. We must encourage risk taking across all levels in the organization. It is leadership's responsibility to create, cultivate, and reinforce this safe environment. Small actions can undermine this safety if we are not intentional in our efforts. A safe environment is the foundation for taking risks and sharing possibly conflicting ideas. We cannot empower a workforce to be innovative if the employees feel they are operating in an unsafe environment.

The blame game squelches innovation and causes people to go on the defensive. Julianne M. Morath, former chief operating officer at Children's Hospitals and Clinics of Minneapolis, used new language to encourage "blameless reporting." She wanted to remove terms with a negative connotation. Morath preferred to see terms such as "investigating" replaced by less negative terms such as "studying." Harvard Business School professor Amy C. Edmondson affirmed that "by avoiding words that implied blame and encouraging language conducive to learning from failures, Morath was trying to make it psychologically safe to talk about error."[3]

Spotlight on Innovation
"Learning from failure is a hallmark of innovative companies."[4]

Leaders cannot shoot the messenger when he or she comes forward with news of failures. A safe environment is one where people do not have to be concerned with self-protection. There is no shame in intelligent failures. We cannot learn from failure if we are focused on distancing ourselves from the failure, hiding it, explaining it away, and protecting ourselves. A safe environment allows us to admit our failures to then learn from them. As Edmondson shared, "Eli Lilly throws failure parties to celebrate clinical trials or scientific programs that were intelligent but that nonetheless failed."[5]

Granted, it is not easy to reframe our thoughts about failure. But it is possible. When Alan Mulally joined Ford as CEO, he used a color-coded reporting system for his managers to identify failure.[6] Reports were presented on green, yellow, and red paper to indicate their status—with red for problems or failures. Managers in early meetings coded everything on green paper indicating "good" status. After Mulally challenged the managers, a lone yellow report finally showed up—to receive applause from him. Learning this was a safe environment in which to share bad news, subsequent meetings then saw plenty of reports on yellow and red color paper.

As leaders cultivate this safe environment, it helps us overcome our inhibitions. We feel safe being vulnerable. This encourages people, then, to really use their imaginations without worrying about what others think. We do not fear that others will consider us stupid or even make fun of us as we share our wild ideas. Without this safety, however, these inhibitions cause us to censor our ideas and innovation suffers.

Procter & Gamble understands the importance of admitting our failures and making them public with no blaming. The company has taken this to the next level by creating the President's Fail Forward Award. The award is presented to the "team or individual that enabled the organization to significantly learn from a failure and as a consequence enables a future project or team to move forward much faster and/or better."[7] Edmondson suggested that an environment of psychological safety exists "when people feel able to express ideas, ask questions, quickly acknowledge mistakes, and raise concerns about the project early and often."[8]

Furthermore, in an unsafe environment, there is a concern that sharing their knowledge can make employees vulnerable and less valuable to the organization. In a safe environment we know our value increases as we contribute our knowledge and expertise to generate ideas and new solutions.

A safe environment, then, is based on an environment of trust. This encourages people to get to know one another and develop strong relationships. Trust is essential when conflict and disagreements arise. Because innovation is fed from a diversity of talent and ideas, this trusting environment enables us to disagree with one another and to effectively manage the conflict inherent in this diversity.

Spotlight on Innovation

"As a leader, you can make or break your organization's ability to support and sustain innovation."[9]

Leaders can help us learn to manage this conflict without allowing it to become personal. Process guidelines assist us in addressing our emotions and bringing our differences to the surface to discuss them in a calm way. If we have not invested in building these trusting relationships with others, we will not be able to weather our disagreements.

When leaders create a safe environment, people are more likely to admit failure, discuss it, learn from it, and move on. (The importance of failing fast was discussed in Chapter 2.) We can fail faster if we avoid blaming. We might say we can expedite the failing process! A sense of urgency in recognizing when we have failed helps us keep the momentum going and provides faster cycle times. Declaring failure early in the process or otherwise failing fast frees up our resources, so we are able to divert them to our next project. If we reward failure, there is less of a tendency to fall into the trap of escalating commitment—where we prolong the failure and spend more resources trying unsuccessfully to make it work.

Walking the Talk to Reinforce the Culture: Modeling the Behaviors We Value

To nurture an innovative culture, managers must model the appropriate behaviors. To reinforce culture, we need to focus on what members of management do—the behaviors they engage in—more than on what they say.

Terry M. Farmer and Xavier Butte, co-founders of EiQ, described the effective role model in an innovative culture as one who "asked odd questions; listened effectively; was curious and open-minded; and shared personal stories of mistakes, failure, and learning."[10] Only by listening to others with contrary viewpoints can we generate innovative solutions. We need to truly embrace different perspectives. These model leaders also collaborate with those in and outside the organization. This may mean asking newcomers and those who are closest to the work for input. The message being sent by these behaviors is that it is okay not to know everything. The organization's leadership has to create this safe environment—modeling the behaviors—asking questions and admitting they need advice and ideas from others to communicate to all employees that this behavior is acceptable throughout the organization. The safe environment allows people to be vulnerable enough to ask those odd questions. Open communication, then, encourages a "culture of questioning."

Rethinking Control

Just as the innovative culture must be adaptable and flexible, our control must also be flexible. Though it might sound like an oxymoron, flexible control is critical. In heeding Hamel's advice to challenge today's approach to management, we need to examine our control-oriented approach to managing people, processes, and resources. Hamel recommended that we challenge our assumptions.[11] We might ask if we really need to do things this way. Why? Is there a better way? If we were starting from scratch, is this the way we would do it?

Leaders may also want to consider the language they use. Daniel H. Pink advised replacing controlling language such as "must" or "should" with words such as "consider" or "think about."[12] These are more noncontrolling in nature.

Spotlight on Innovation

"Excellence in leading innovation has far less to do with the leader having innovative ideas; it has everything to do with how that leader creates a culture where innovation and creativity thrives in every corner."[13]

The traditional view of control (versus freedom) in management should be reconsidered. Freedom is empowering! W. L. Gore has no formal hierarchy—yet it still has discipline with this freedom. Those organizations with strong cultures often find that they require fewer rules and regulations because the values of the organization that permeate all levels and are embraced by everyone guide the behaviors of the organizational members (and naturally provide that control). This element of a strong culture was one of the characteristics ("loose/tight") of excellent companies proposed by Tom Peters and Robert Waterman in the classic management book *In Search of Excellence*.[14]

Pink was quite clear, though, that "encouraging autonomy doesn't mean discouraging accountability."[15] Another example is provided by the company Atlassian with its FedEx days. Employees are encouraged to take days off from their normal work to address other problems or projects. But then they are required to deliver results the next day. There *is* autonomy with accountability!

Pink shared an example of autonomy: "At Georgetown University Hospital in Washington, D.C., for instance, many nurses have the freedom to conduct their own research projects, which in turn have changed a number of the hospital's programs and policies. Autonomy measures can work in a range of fields—and offer a promising source for innovations and even institutional reforms."[16]

Bureaucracy and its red tape often get in the way of innovation. As we rethink our need for control, we may want to consider more flexible approaches. For example, we do not want to use the same metrics for all our business units; established and entrepreneurial ventures may warrant different metrics. One-size-fits-all measures are no longer appropriate. We should rethink metrics to consider the impact on innovation. Measuring all our businesses by the same metrics can discourage innovation, especially in new businesses. Traditional organizational processes and metrics may work for established/existing businesses, but newly launched businesses often require new processes, approaches, and metrics. (This also explains why some businesses spin off their new ventures to isolate them from the traditional business.)

Spotlight on Innovation

"One of the most prized corporate attributes is innovation. But many of the natural byproducts of corporate success—hierarchy, routinization, the elimination of risk—can stifle innovation."[17]

Flexible planning becomes essential. Whereas at first glance, flexible planning may seem to be yet another oxymoron (and one that cannot be implemented), it is actually a requisite component of the innovative culture. For example, we should consider the need to have flexible funds available. If an idea bubbles to the surface, we then have the option to devote resources to its development immediately. We cannot necessarily plan ahead for this. But we *can* plan flexibly enough to leave ourselves some wiggle room—and discretionary funds.

Flexible work arrangements are one of the most effective tools for companies to engage their workforces. Freedom is the foundation of the results-only work environment (known by the acronym ROWE). Best Buy has effectively used the ROWE model since 2005. In an article about ROWE, one Best Buy employee said the model "is one where every employee is free to do whatever they want, whenever they want, as long as the work gets done."[18] Earned vacation days are eliminated; employees are trusted to take the time off that they need without a hard and fast accounting system for tracking it.

ROWE is based on the results produced, and these are measurable results by which individuals can be held accountable. This environment moves away from the traditional measurement of face time (or hours worked). Yes, ROWE may strike fear in the hearts of die-hard control managers. And although we may not need to go to this extreme, there is definite merit demonstrated by providing more freedom to our workforces. Rethinking control, then, is based on a foundation of trust.

As leaders, providing people with autonomy is essential to the innovation process. Yet it is also necessary to ensure there is a clear understanding of the goals—to serve as the direction. Unfortunately, we cannot dictate innovation, but we *can* set goals. Once individuals know the target (goal) is clearly in their sights, there is no need to micromanage individuals; instead, managers need to get out of their way. Management's role is to emphasize the desired outcomes—not *how* they should be accomplished. Letting people decide how they will attain those results/outcomes is providing autonomy. McKnight's philosophy at 3M was to "hire good people, and leave them alone."[19] The workforce in an innovative culture needs freedom to innovate. Just like a fine racehorse must be given its head to run a winning race, our workforce members must be provided freedom to win their race to innovation.

Spotlight on Innovation

"Chaos is an essential part of innovation. Let the front end of the process be unstructured and very informal—even different for each part of the organization, if necessary. Begin using metrics only after a consensus exists for what is important to measure."[20]

A "Space" in Our Workplaces for Innovation: Place, Time, and Tools

In managing people within an innovative culture, we should give consideration to the "space" needed. Leadership needs to encourage collaboration since innovation is seldom generated in isolation. Effective collaboration requires a physical (or virtual) place, space in our schedule (or time), and the tools to leverage these.

Place

The "cross-pollination" of ideas with cross-functional teams that brings together different perspectives is only possible with a forum for sharing ideas. Creating forums for this collaboration to share ideas is critical—and requires deliberate attention. We cannot assume it will happen on its own, though at times, it may.

If we agree that everyone must be innovative (versus creating pockets of innovation), then we must provide forums available to everyone. Boundary-spanning networks aid in this collaboration process. Professor Rob Cross and Chris Ernst, vice president of leadership and organization effectiveness at Juniper Networks, provided a broad definition of these boundaries. They include vertical and horizontal organizational structures, stakeholder groups, demographic groups, and geographic location.[19] Innovation communities allow people with mutual interests to connect with others to share ideas and to communicate openly. These communities create the potential for action of a collective nature. The successful cross-pollination of ideas needs close connections both vertically and horizontally— that is, between departments and hierarchical levels.

An example of a best practice is provided by a case study of a health care organization that discovered the importance of creating a forum for innovators to connect. They held a collaboration conference and invited various stakeholders (including doctors and nurses) to explore customer service innovations. Patients and caregivers were also invited on field trips to visit other health care facilities. In addition, individuals were allowed the opportunity to work temporarily in other areas to learn more. A key lesson from this organization is the value of tackling one aspect of the whole at a time. For instance, this health care organization did not try to change the entire facility, but rather focused on customer service. And it certainly leveraged its ability to create a forum for the cross-pollination of ideas.

Juniper Networks held an offsite event called the Innovation Challenge that spanned three days. The company found that location matters. It traded in the traditional offsite conference rooms for a garage to hold the meetings. It traded in conference opening remarks by executives for teams hitting the streets to ask what people thought about the firm's technology and its industry. It traded in PowerPoint presentations for rolls of butcher paper. And it traded in one-way top-down communication by company executives for two-way brainstorming with experts across functional lines. The event not only generated innovative ideas, but it sent an important message about the environment of collaboration. When we want new outcomes, perhaps we need to think in terms of new forums (and new "spaces") to produce those outcomes.

Spotlight on Innovation

"Employers across industries have embraced creative open floor plan offices as a way to convey their culture and attract fresh talent. Designs may range from Zen to zany, but the common goal of these spaces is to spur creativity and collaboration."[22]

There is no perfect method in designing effective workspaces. Instead, we must align our spaces with our unique environments—both for the organizations and the workforces. Janet Pogue, with the large workplace design firm Gensler, is reported to have said that "choice is especially important for the growing number of knowledge workers who are more apt to expect workspace options to foster engagement and productivity."[23] Recognizing the importance of physical surroundings in creating the environment that allows innovation to flourish, Trenam Kemker, a Florida law firm, uses "collaboration zones" to encourage personal interactions.

GlaxoSmithKline in Philadelphia uses shared workspaces to facilitate collaboration and to enhance innovation. Layout design can improve and change the way people interact and the way people communicate to share ideas, and ultimately, to innovate. The trend has been toward designing these smaller work areas that promote collaboration.

Open, flexible office spaces are popular today in nurturing an innovative environment. GlaxoSmithKline uses a "neighborhood" concept to change the way people interact. The departments have eliminated walls so that people can see each other. This open office concept has reduced the number of e-mails sent internally, thereby expediting the communication process.

The open office design provides space for people to come together to collaborate in a relaxed environment. Whiteboards can be located throughout the area for capturing ideas, and "nooks" can be added to provide smaller areas for more intimate meetings or to serve as a space for quiet reflection. A variety in workspaces has enabled companies to match their cultures and offer nontraditional spaces for workers. Game rooms have been used to allow employees to connect on a personal level. The Lending Club in San Francisco even houses a miniature golf course in its offices. IDEO, the global design consultancy known for its innovation, has designed its workspaces with play in mind. It emphasizes teamwork and fun by designing creative rooms. Attention to both space and time are critical.

Time

Action and reflection are required for innovation. Though we are a society generally geared toward doing, we must provide time for everyone to engage in thoughtful reflection to effectively foster an innovative culture. Time is necessary for people to innovate. Companies known for their innovative cultures (such as 3M) often build in time for reflection by providing freedom from "normal" activities.

A concerted effort to spend time reflecting is needed. Otherwise, people will find themselves caught in the doing trap. We also need time to reflect on our failures. Once failure occurs, it must be analyzed for root causes. Pixar uses a process whereby project members identify what they would repeat and, equally important, what they would not repeat moving forward.

Spotlight on Innovation

"HR practitioners said management style is overwhelmingly the biggest obstacle to effectively initiating workplace culture change."[24]

Tools

A space for innovation is also about providing innovators with the tools they need. Contrary to popular belief, our problem with innovation is not a shortage of ideas. We need an "innovation engine"— an idea-management process—which must be built by management. That is, we need a framework for capturing these ideas; otherwise, they may be lost, and we could forego wonderful opportunities for innovation. In some organizations this process has been an internal discussion board where ideas can be generated and shared.[25] The key is for organizations to identify the criteria for reviewing ideas and how the ideas are selected for implementation and further development. Additional tools can then be selected around those criteria.

Another example of a best practice is the use of a scorecard. This can be used to track ideas. The scorecard might include the number of ideas and even the source of the ideas. The status of each of these ideas is monitored along with the financial impact—both costs and benefits. Nonfinancial benefits are tracked on the scorecard as well.

"Gallery walks" can be considered to provide a visual depiction of the progression of a group's thinking. And to widen the number of people who can participate in brainstorming, we can leverage technology to provide virtual brainstorming sessions with employees who are geographically dispersed.[26]

The tools used may not be as important as how they are implemented. Management must communicate the process broadly; when people have an idea, they need to know how and where to present it. And they need to be heard. That is, some response or follow-up should be built into any program or approach.

Who Should Lead?

Leadership is responsible for creating the foundation on which the innovative culture is built. That is, leadership must articulate the focus on the importance of innovation as a core value, model the behaviors desired in others, and maintain momentum for innovation.

The role of leadership is central to the innovative culture. Farmer and Butte opened their article, "Inspire to Innovate," with a powerful question: "What role does a leader play in the innovation process? The short answer is: a big one. Unfortunately, research continues to highlight leadership as a top barrier to innovation."[27] The authors went on to suggest that leaders have an opportunity to promote innovation through their actions. The good news is that there is an opportunity, then, for leaders to change this.

A consideration in the management of people is thinking in terms of the roles that help lead the innovation process. That is, we should consider the "connectors." These are the informal roles identified in the alignment of teams. The "cross-boundary broker," as identified by Cross and Ernst, "brokers" innovation across departments.[28] The individuals in these roles are often seen as the "go-to" people for attaining information and assistance in problem-solving. Perhaps most essential is the role of "energizers." These individuals inspire people and serve a pivotal role in engaging people.

To effectively manage an innovative workforce, we should consider leaders who have imagination and a natural curiosity. They will be more likely to be open to a variety of perspectives and to ask thoughtful questions to maintain the innovative momentum. Those who take a more participatory approach to the management of people are better suited to the open environment we are cultivating.

Spotlight on Innovation

We should "advocate environments that encourage risk-taking, the free exchange of ideas, legitimate conflict, active participation, and the use of intrinsic rather than extrinsic rewards."[29]

We may want to think nontraditionally when seeking managers to lead innovation projects. Those who have excelled in current (successful) businesses may not be the ones to lead new ventures. A different skill set is required. A better approach might be to assess the leader's competencies across the skill set identified for success in an innovative environment.

Successful leaders in innovative cultures recognize that trust cuts both ways. They trust their employees and run interference for them. Employees in turn know their leaders have their backs and create that safe environment for them to experiment. The most effective managers are comfortable not being in the limelight, freely give recognition to others when earned, and are comfortable giving up control. They are managers confident enough to empower others and even provide employees the opportunity to champion their ideas outside their own departments. These managers want to develop others and excel in the coaching roles. And they are good team players.

Innovative leaders go beyond an open door policy. Their open communication style makes them supportive co-workers. They are confident in both what they know and what they do not know. And they are open to sharing what they know, never hoarding information from others. They are skilled in providing feedback and can effectively listen to receive valuable feedback—and then act on it.

Managing people in an innovative culture requires the trust to take risks and the flexibility to change paths quickly with the discipline to learn fast and move forward. Essential to management is diversity of functional background and expertise. Just as our innovative cultures need diverse workforces, we also need diverse management teams with an ability to manage cross-functional teams.

What Leaders Should Measure

Given that we get what we measure, we should consider rethinking what we measure. People focus their efforts on what gets measured. That is, they focus their actions on these "measured" behaviors. Traditionally, organizations have focused on efficiency versus innovation—and have likewise used efficiency measures. Yet these are not the measurements that foster innovative behaviors. The process of measurement is important to innovation and therefore must be integrated into the evaluation system.

Effective leaders in an innovative culture must focus on what they say is important. The performance appraisal process, then, must include goals related to innovation. We cannot say that innovation is valuable and then fail to "measure" it. The performance appraisal interviews should be used to open additional dialogue around innovation and the company's expectations in regard to employee performance.

Spotlight on Innovation

"Executives and managers must influence, convince, and sell employees rather than order them around. Some even fear the approach because it requires a completely new executive skill set and a great deal of patience."[30]

We must rely on managers and leaders throughout the organization to have these discussions. First-line managers are particularly crucial in supporting the innovative culture with attention to these discussions—by virtue of the fact that they are the front line!

This focus begins with including innovation in the job description. In addition, professional development needs to include innovation skills (which were discussed in Chapter 4). We might also consider building questions about failure into the performance appraisal process.

An appropriate reward and recognition system is essential in communicating that the organization appreciates and values innovation. This can also serve as a motivator (as discussed in Chapter 5). However, it takes more than just creating the program. Leaders must model innovative behaviors.

Chapter 7

The Shoemaker's Children: Is HR's House Innovating?

The new view of innovation applies to products, processes, and people. It is about creating an infrastructure and developing the capabilities for the workforce to engage in innovative behaviors. This has HR's name all over it! It provides an opportunity for reinventing others' jobs and our own. As innovation has been identified as a critical competency across most organizations today, it is our responsibility in HR to take ownership to ensure this competency is developed in our workforces. Of course we cannot lead the charge if we ourselves are not walking the innovative talk.

As HR professionals we are responsible for reinforcing our cultures of innovation. We might say that innovation begins at home. The question posed in this chapter, then, is whether we are providing a model for the rest of the organization and whether we are aligning our processes to foster a culture of innovation. Some of the innovative approaches that can be considered within HR are explored, and some initiatives from other organizations are shared.

The Call to Action

Keith Hammonds, deputy editor at *Fast Company*, issued a call to action for all HR professionals with his well-publicized article, "Why We Hate HR."[1] The tongue-in-cheek approach reminded us that HR needs to be less administrative and more strategic. And Hammonds is not the only one issuing a rally cry. The University of Southern California's Center for Effective Organizations has studied the HR profession's ability (or perhaps more accurately, *inability*) to keep pace with the changing landscape. Bottom line, the result is that HR is lagging and has further been characterized as "slow" in meeting our challenges.

Spotlight on Innovation

"Being at the table obligates us to think outside our own organizational boxes and to participate in the interests of all stakeholders."[2]

Perhaps of even more concern, the Center for Effective Organizations reported that HR was no more of a strategic business partner in 2013 than in 1995! Yet there is good news. Fostering an innovative culture is more aligned with the strategic approach, and as HR professionals learn how to assist their companies in espousing innovation, we automatically become better strategic business partners.

As the business environment shifts to a focus on recovery, the need for innovation has never been greater. And this challenge extends to our HR departments. Reengineering processes within the HR department were cited as a key initiative for HR professionals in a Towers Watson 2014 survey as reported by the Association for Talent Development.[3]

William Sebra, president of Futurestep's North America business, is unequivocal in his message:

> The importance of HR innovation to employees can no longer be underestimated. Globally, the HR industry is changing and the North American employment market, viewed as one of the most exciting and diverse globally, is also at this crossroads. In a market as competitive as the U.S., securing and keeping the best talent is crucial. Candidates and professionals overwhelmingly agree that more needs to be done to incorporate innovation, not only to retain the best talent but to keep motivation and productivity high.[4]

There can be no more "business as usual" for HR. Founder and CEO of Human Capital Source, Jac Fitz-enz suggested that "perhaps most encouraging is, amid all that is exciting and emerging in the field of HR, the function is still coming of age and has a lot of room to innovate."[5] Now is the time to answer the call for what Professors John Boudreau and Ed Lawler call a "new wave of HR professionals."[6] We are poised at a crossroads with tremendous opportunities if we simply heed the call and accept—and perhaps even embrace—the challenge to foster an innovative culture by leading the charge. And this means leading by example!

Wanted: A New Breed of HR Professionals

This is a new dawn and a new day for all of us as HR professionals. A transformation has been underway as we have addressed new challenges in a rapidly changing and hypercompetitive business landscape. The call today is for agile HR professionals who can rethink our traditional (and perhaps outdated) business models.

The silo mentality of promoting from within the HR unit contributes to many of Hammond's points about HR's less than stellar perception by those outside the field. It is time to consider facing the future with HR leaders who have been educated in fields outside HR—and who have experience outside the field of HR. Though it might sound like heresy, it is the antidote for inbreeding and the key to an understanding of the big picture and the interconnectedness of the business as a whole.

Spotlight on Innovation

"One thing to note is the strategic role that HR plays in creating and facilitating a culture of innovation. Hiring, training, cultural onboarding, recognition and rewards, even health and well-being programs are critical to attracting and retaining.[7]

Andrew McIlvaine related the challenge at Tesla: "They also want to create an HR department that's as innovative as the company it serves."[8] It is difficult to effectively lead a charge for innovation throughout the organization if our own department lacks this very innovation.

A 2011 survey of global leaders in human resources conducted by IBM reported their number one challenge to be driving organizational innovation.[9] There was a major disconnect in the execution of this challenge since only about half of these leaders reported following through to actually take any action. Whereas 70 percent of the respondents reported a significant role played by HR in guiding innovation, most do not have the necessary tools to recruit creative applicants. A little over half do not align their performance management systems with innovation. It is time for us as HR professionals to ensure we model the behaviors desired—and begin our commitment to innovation at home.

Thinking a Little Differently: A Taste of Our Own "Medicine"

It is time to rethink and reframe our work. We no longer have the option of maintaining the status quo. As HR professionals we must add value, but in a new way. And we can do so only with new tool kits. Perhaps it is time to even think of ourselves differently—as business facilitators.

Not only must we reevaluate the bureaucratic, top-down structure, but we must also consider the effectiveness of the traditional silo structure. With agility in mind, some of our colleagues in other HR departments have begun the shift away from the "physical housing" of the HR department all together. More progressive firms are moving HR professionals to be located physically with the businesses they are supporting. This certainly helps to better understand the needs of our customers when we are sitting among them.

Spotlight on Innovation

"When talent managers design programs, forget the structures of the past and rethink the purpose of each program . . . . On training and development, people want easy access to learning resources."[10]

One approach to thinking more creatively suggests that we reverse our thinking. Some HR departments have done just that in the attempt to reframe exit interviews. When we consider the goal of exit interviews, it is usually to understand why people are leaving—and then to stop others from leaving for similar reasons. We are attempting to get to the root of the issue for organizational voluntary terminations. Yet, at the point of the exit interview, the decision has already been made for the individual to leave the organization.

Sharon Jordan-Evans and Beverly Kaye, co-authors of the article "More Stay Interviews, Fewer Exit Interviews," counseled using stay interviews, which is reversing our thinking.[11] Their advice was to ask, "What will keep you here?," remembering that this is not a one-time event. This question should then be paired with "What about your job makes you jump out of bed in the morning?" and "What makes you hit the snooze button?" These questions get at the heart of what excites people (and retains them) and what we might address to "save" them. This is a total reversal in our thinking—and an innovative approach to understanding turnover.

A number of companies have begun to reverse their thinking on formalized tracking systems and handbooks. Historically, companies have diligently tracked the number of vacation hours earned and used by their salaried employees. Some more progressive companies have communicated their trust in employees by turning to ROWE, or results-only work environment (see Chapter 6), and allowing them to take what is considered an appropriate amount of time off. While guidelines and expectations of responsible behavior are provided, a culture of trust communicates that employees will keep the best interests of the company in mind. This practice has also been expanded to include expense policies for some companies. Perhaps the reversal in thinking is based on trusting employees versus monitoring them. Netflix is one company that has used an honor system for employee vacation days, holidays, and sick days.

Companies thinking out of the box and reversing their thinking have also begun to embrace the "anti-handbook" handbook. The trend for years was toward adding longer, more detailed policies—and more pages—to our handbooks. A reversal in thinking has HR departments creating more informal and briefer handbooks. When describing the handbook that she rolled out at Yahoo, Libby Sartain said she would not put "dumb" policies in the book.[12] If someone was doing something wrong (such as skateboarding down the hallway), that individual's manager would be responsible for pointing out that the behavior was inappropriate rather than having HR develop a policy for the entire workforce. This again builds on the culture of trust—so fundamental to fostering an innovative culture.

It is easy to get caught in the trap of doing things the same way over and over. This is the antithesis to innovation. As role models, then, rethinking the status quo is essential in our own HR function—in all we do. The recruitment function at the Boston Red Sox organization was reframed to leverage a digital approach to the process. Some companies have recreated training by using QR codes to deliver on-the-job training (by scanning) when needed or bite-sized, mobile resources available on demand.

Spotlight on Innovation

"The nature of HR itself demands that organizations develop new capabilities and that HR's role is to reevaluate its competencies and develop new ones to help in the overall strategic redesign of organizations."[13]

These examples focus on the autonomy and freedom that are extended to employees through rethinking the approach to policies in the HR department. Employees need room to be innovative—that is, they need freedom. Yet this certainly does not mean ridding the organization of all policies. It just means evaluating the broad guidelines and parameters for the organization's workforce rather than providing the minute detail of all their interactions with the company.

Providing freedom in how employees' work is to be accomplished enhances their intrinsic motivation—so critical to creative individuals. Freedom also allows individuals to own their work and use their skills in ways that we may not have prescribed.

HR's Own House: Some Ideas to Ponder

The potential for competitive advantage exists through the use of more innovative practices in human resources. The talent management strategies of the past are no longer sufficient to meet the challenges of today's environment. In fact, opportunities abound in this area. The performance appraisal process is in dire need of overhauling. Our hiring processes in some cases may be antiquated and focused on the short term (versus thinking of hiring for the future), and opportunities exist to consider new approaches for leveraging technology, especially in more effectively identifying candidates and onboarding them.

Borrowing ideas from other functions (and even other companies and industries) requires that we simply open our minds and change our perspectives. In the IT field, high-tech professionals have successfully used "hackathons" as a more current approach to brainstorming.[14] This idea is now taking root in other functions—including HR. A hackathon was sponsored in early 2014 in San Francisco to generate ideas around the employee engagement challenge. The event enabled diverse people to collaborate to leverage different expertise and to explore the cross-pollination of ideas. This can be an important tool for cooperation and collaboration in getting everyone involved.

Hackathons have also been used in the college recruiting process. These IT competitions enable candidates to highlight their skills and abilities for prospective employers. Students at Rochester Institute of Technology (RIT) have been successfully recruited by firms such as Facebook, Toyota, and American Greetings. Events such as these allow companies to gain insight into students' performance, and students gain insight into the company culture.

The HR department has been highlighted at 3M for its role in recruiting, performance evaluation, and compensation. Fostering an innovative culture requires that HR pay particular attention to hiring for innovation. Recruiters must seek out creative types who are inquisitive and open when hiring for their cultures. (Hiring for innovation was discussed in more detail in Chapter 3.) Once innovators are hired, we must provide training opportunities and design rewards for innovation. As HR professionals, then, we can partner with managers to provide training in creative thinking skills, brainstorming, and problem-solving. That is, we can provide the tools required to support innovative processes and the development and growth of our workforces that are critical in an innovative culture. HR can commit to building expertise by also funding appropriate seminars and conferences.

Spotlight on Innovation
We must be aware of "the need of the organization to import new skill sets; attract, nurture, and retain the right professionals; and, more important, to keep them motivated for the right challenges to transform employees into a winning team."[15]

So what else *can* HR do? We can emphasize programs that feed the innovative culture and enhance the engagement of the workforce. Harvard Business School professor Teresa M. Amabile suggested that "of all the things managers can do to stimulate creativity, perhaps the most efficacious is the deceptively simple task of matching people with the right assignments."[16] HR needs to lead the way in helping the organization rethink how we view talent—and how we recruit. HR plays a big role here to ensure managers have the necessary information about employees and jobs to assist in this matching. We might also consider lateral career moves to align people and jobs.

Everyone has a responsibility to support the process of innovation. HR is not exempt. In fact, HR's role is perhaps greater than any other in the organization due to its central position. HR touches everyone through its processes. We have seen that innovation is affected by recruitment, job design, performance appraisals, compensation, and development: all functions of HR. But perhaps our greatest challenge is rethinking our own jobs (to recreate them) and our roles in promoting an innovative culture.

We established in Chapter 3 that attracting and retaining employees is a critical function of HR in facilitating our innovative cultures. Attention to recruitment and selection (talent management) begins with an evaluation of our job descriptions. We may want to consider thinking in terms of emphasizing accomplishments and results. And, as in the case of the hackathons used at RIT, we might further explore recruiting talented individuals while they are still students!

A commitment to innovation must be integrated into the performance appraisal and compensation process. A focus on organizational goals in performance management is vital.

We must encourage line managers to have conversations around those goals more often than just once a year. Zappos has traded in its traditional once-a-year performance appraisal rating process for a system of regular feedback focused on the employees' performance around the company's core values. The new system is used solely for developmental purposes rather than for administrative purposes.

In addition, management must move beyond a focus on just performance to emphasize coaching and developing. When Google realized that its performance management system was simply not working, it identified the "Eight Habits of Highly Effective Google Managers" from performance appraisals of "good bosses."[17] HR then began coaching managers around these leadership skills.

Organizations have shifted away from managing by command and control, a positive trend given that the previous approach ran counter to the innovative culture we want to cultivate. There is no place for dictating all decisions in the innovative culture. Instead, the role of management should be more like facilitators and coaches. As HR professionals, we are pivotal in preparing our workforces for this shift by creating the infrastructure and developing the competencies to support them.

Spotlight on Innovation

"At innovation-driven companies such as Qualcomm and LinkedIn, these [innovation] competencies are baked into the way people are recruited, trained, and managed."[18]

Keepers of the Culture

Michael Stanleigh, president and CEO of Business Improvement Architects, clearly grasps the reach of HR in fostering an innovative culture. Using a well-known company as an example, he stated,

> Apple has about 35,000 permanent employees, yet continues to retain a culture of innovation through their HR practices. They hire, reward, and recognize employees for a common desire, energy, and enthusiasm to create great products. They encourage employees not to be afraid to fail. There is no punishment for this.[19]

As we can see with this example, we do not have to be a small company to be innovative. But we do have to nurture our culture. And as HR professionals, we have to assume the role of "keeper of the culture."

Reinforcing the culture can be achieved in numerous ways. Paying attention to the people we hire and socializing them to our culture are two ways. The stories and artifacts of culture can also be preserved and institutionalized. Nike uses an old Winnebago as a conference room—playing on the legend of Phil Knight. The co-founder of Nike was regaled for selling the company's first pair of shoes in a similar RV.

Preserving artifacts reminds us what we value. This is embraced with the infamous waffle iron at Nike that remains on display. (Legend has it that track and field coach and cofounder of Nike Bill Bowerman destroyed it while creating the now famous rubber soles on Nike shoes.) And of course, the classic innovative culture story known to those within and outside 3M is the beginnings of the Post-It note as the "failed" glue. As HR professionals we can think about the kinds of stories and artifacts we can preserve from our own cultures—perhaps in a unique way.

When we hire for culture fit, talent retention can be an important by-product in a strong, cohesive organizational culture. Nike refers to employees with the company for less than 10 years as its rookies. In strong organizational cultures, the values are so deeply ingrained that people know them like their own children's names! These deeply ingrained values guide us in our behavior.

Reinforcing the culture also involves recognizing what areas may be considered "hands-off" and essential to our business. For example, UPS recognizes how critical its drivers are to its success. High turnover was experienced due to the physical element of loading the trucks. Therefore, the company created new part-time positions to load trucks, thus allowing drivers to do their important work; this solution reduced turnover.

Spotlight on Innovation

"Learning professionals recognize they are in pivotal positions to act as change agents, helping their executives and workforces acquire the skills needed to navigate change more successfully."[20]

If we want decentralized environments where employees are empowered to work in new ways, we must create the systems and processes to support them. Regular training and developmental opportunities should be considered in the areas of communication, critical thinking, teamwork, creativity, leadership, managing innovative teams, and change management for the workforce across all functions (including our own).

As keepers of the culture, we also want to consider training in the culture. This begins with the recruitment process as we communicate to applicants who we are and what we stand for. The onboarding and orientation of new employees should also focus on key elements of our cultures to socialize our newcomers. But the training should not end there. First-line managers are principal components in reinforcing our culture by "living it" and serving as role models. They, too, should have refreshers in culture training.

Whereas we might ask our applicants what they have tried new lately and what they learned from it, the question might be equally applicable to ourselves. What new things are we trying in HR? What is working? What is not working? And most importantly, what are we learning from these? If we have not tried anything new lately, why not? And are we missing an incredible opportunity to help nurture our innovative culture. Or worse, we could be hindering our organizations' ability to foster an innovative culture!

Chapter 8

Parting Thoughts: Putting It All Together

Although culture drives innovation, there is no specific, detailed blueprint or step-by-step manual in fostering an innovative culture. As suggested in the article "Drive an Innovative Culture," "an innovative culture is not a one-type-fits-all approach."[1] The most effective cultures will be those that embrace basic characteristics supporting innovation, but do so within the uniqueness of their own organizations and values.

Fostering a culture of innovation requires the alignment of a large number of elements, carefully orchestrating all the pieces to fit together. Each element is part of a moving system that ultimately creates the culture where innovation can thrive. So we must consider that every innovative culture is different—though with some key, shared common elements. As HR professionals, we are one of these distinctive elements of our culture. Just as we should assess how the organization is doing in moving toward the innovative culture, we want to assess ourselves in our roles as contributing to developing this culture.

Spotlight on Innovation

"A wide variety of key components of a company must be brought into alignment with the goal of promoting innovative behavior. Concepts must guide the process of reshaping the work setting so that it continuously stimulates and supports creativity, risk taking, and learning from mistakes—all necessary behaviors for innovation to occur."[2]

Our Own Special Sauce

Each chapter in this book discusses different elements of nurturing an innovative culture, but how those pieces come together in the end is what matters. Addressing only one, or even a few, of these elements will not successfully guide a culture of innovation. Like a puzzle, our culture is not complete (and fully functioning) without all the pieces. And it certainly will not be sustainable over the long term if it is not complete.

All the elements we have discussed are dynamically linked. Each organization simply adds its own twist—or "special sauce"—to make it its own. We put the elements together in a way that reflects our own organization and our own approach to innovation. There is no one right way to form a culture that welcomes innovation. That is, there is no cookbook recipe.

There are, however, key elements to be addressed in our own ways. These were the focus throughout the book. Fostering a culture of innovation means innovation is highly valued across the organization—and woven into the fabric or the DNA of everything we do. It is not a part-time initiative or a once-a-year agenda item. When effective, it becomes a way of life for us. Literally everyone in the organization must be responsible for innovation.

Scott Anthony, managing partner of Innosight, asserted that innovation extends beyond the traditional thoughts of technology that disrupts entire industries or revolutionary products.[3] Today's innovation includes improvements of any size—otherwise described as "something different that has impact." It does not necessarily have to be about new knowledge. We can leverage the knowledge currently in existence, but in new ways. This approach opens the door, then, for everyone everywhere to innovate.

And the good news is that many members of our workforces are creative; we just need to awaken that thinking with a culture that encourages them. Our role as HR professionals is critical in running interference and removing barriers for people. This requires that we understand what these barriers are likely to look like.

Embracing a Change in Culture: Moving from the Known

Fostering a culture of innovation requires making changes across the organization. We cannot disrupt our businesses if we continue going down the same path. Yes, disruption is good. In strategy, we refer to creative destruction—with Apple being cited as a company that creatively "destroys" its own products. Doing more of the same simply will not work. Therefore, we must overcome myopia and avoid inertia. Bureaucratic fossilization results when we institutionalize our ways and fail to change.[4]

Innovation requires a culture that can sustain itself. It is not about doing more of the same, but rather about changing and embracing new ways of doing things—even when the old ways are still working. It is disrupting our own success.

The seeds of our success may beget the seeds of our downfall. Our past and current successes often blind us to new ways of doing things. Continuing to do things the same way closes the door to considering other options. This inability to innovate ultimately means we fall behind. We can see this played out in the battle of two big-box retailers. There are compelling lessons to be learned in the stories of Best Buy versus Circuit City. Circuit City reveled in its 50-plus years of success and sat on its 20-year-old point-of-sale (POS) system. Best Buy opted to upgrade its POS system and its business model to a less expensive environment of customers shopping by themselves. When Circuit City finally realized it needed to take a new approach, it was too late, and the company story ended in bankruptcy. The impact of being wed to the same ways of doing things is real—and often far-reaching and negative.

As discussed throughout the book, innovation takes many forms. As in the case with Best Buy, it can mean disrupting our current successful business models. Adaptive cultures can provide a competitive advantage, if we let them. LEGO engineered a company-wide turnaround—all with a focus on innovation. GlaxoSmithKline, the pharmaceutical giant, engaged in innovation by borrowing an approach to analytics used in the Formula 1 race industry.

Keith Jarrett, an American jazz concert pianist, provides us with a great understanding of how innovation can result from less-than-ideal conditions. When arriving late for a concert appearance in Germany in 1975, he found not only the "wrong" piano, but a piano in poor condition. To compensate for the piano's shortfalls, he improvised and delivered a concert recording that became a bestselling album. We can do likewise in our organizations—thereby making lemonade when given lemons!

Recognizing the Barriers to Innovation

Whereas we have discussed the organizational elements that can foster an innovative culture, we should also recognize the barriers that hinder innovation. Some of them may be inadvertent, but they nonetheless can impede our ability to spread innovation throughout the organization. If we know the barriers, we can prepare to overcome them and perhaps even manage around them or make the necessary adjustments.

A negative attitude can stifle our culture and poison the well of innovation. Pessimism can quickly squelch innovation. An innovative culture requires openness and a willingness to take risks. An optimistic outlook helps fuel this mindset. And pessimism has people challenging new approaches with the time-worn excuses we have all heard: "That would never work," or "we tried it, and it didn't work." Pessimism shackles individuals to the known and prohibits movement out of our comfort zones, forcing us to cling to the ways we have always done things.

A lack of openness—in anything we do—can provide a roadblock to innovation. Being closed does not allow new ideas to take seed and grow, essentially shutting down creativity. Being closed to diversity in others fails to allow us to capture different perspectives so critical to the innovative process. Being closed shuts down the communication we need to collaborate with others. And being closed does not build the trust so core to the innovative culture.

Being blinded by what we know hinders innovation. This can include already thinking we know it all (and not trying new things).

Thinking we have already learned everything from a successful track record and continuing to do things the same way closes the door to experimentation. Thinking we tried it once and it did not work may prevent us from revisiting ideas that may now be appropriate. Thinking subject matter experts have all the right answers causes us to miss the point that there may be more than one option. That is, not all problems are solved with one solution. And a fresh perspective from someone outside the field or a novice may yield different answers through a new lens.

Tightly controlling what we do leaves little room for innovation. Tightly scheduling our time leaves no time for reflection and thoughtful analysis. Tightly controlling our structure leaves little opportunity for open communication, collaboration, and the cross-pollination of ideas. Tightly prescribing the behaviors of our workforces through narrowly defined job descriptions and performance appraisals leaves little opportunity for new behaviors. Tightly controlled work arrangements diminish the engagement of employees. Tightly controlled budgets leave no room for discretionary funding for innovative projects and ideas that arise. And tightly addressing the "wrong" rewards gets us what we measure—not the innovation we want.

Whether intentional or not, we sometimes get in our own way. We create roadblocks to innovative thinking with generic, knee-jerk responses such as "we've always done it this way," or "yes, but." These all create barriers to innovation. Our responsibility is to determine how, then, we can run interference and remove some of the hassles to clear the path to innovation. That is, we can consider minimizing control throughout the organization, eliminating some of the hierarchy, and assisting in providing the leadership support needed to cultivate and reinforce the culture of innovation.

Finally, we want to pay attention to any of the organizational elements that inherently run counter to the foundation of an innovative culture. Being risk averse serves as a barrier to experimenting and trying new ideas. A failure to engage all of our employees in the work of innovation is a barrier. Taking a short-term focus keeps our eyes on the wrong ball and sets us up to miss opportunities to innovate.

Spotlight on Innovation

"Innovation springs from the minds of creative individuals working in an environment that spawns and encourages innovation."[5]

Embracing the Paradox of Innovation: Yes, and . . .

Not so ironically, an innovative culture is really about paradox. Even the language of innovation embraces paradox. Harvard Business School professor Rosabeth Moss Kanter suggested, "An organization is more likely to get bigger ideas if it has a wide funnel into which numerous small ideas can be poured."[6] And we have established that to experience more success, we need to risk experiencing more failure—yet another paradox. Failure, then, is a key to success.

Jim Collins and Jerry Porras warned us to avoid the "tyranny of the Or" in favor of the "Genius of the And."[7] Polarized thinking results in choosing between right and wrong. This is based on the polarized assumption that there is one right answer. We need to forego addressing our problems with "either/or" language and solutions to embrace "both/and" solutions. We even need to learn to integrate opposing/paradoxical ideas. Creative thinking requires the simultaneous integration of both systematic analysis and intuition.

Control is often used as the driver in our organizations. Yet this leads to processes that are repeated over and over—the opposite of innovation. The flexible structure advocated for innovation enables scheduling of free time. Google provides its engineers with 20 percent free time for projects. Though serious business issues are addressed, they can be done so with a playful environment that stimulates innovation. IDEO has served as the poster child for playfulness in the work environment.

The innovative culture provides encouragement to fail and even rewards early failure. We want our failures to be fast so we can move on to the next experiment or idea. However, encouraging failures does not mean we abandon our responsibility for them. We still have responsibilities, and people are held accountable. With innovation there are clear deliverables. Our goals provide a sense of purpose, direction, and motivation. They define expected, measurable results. We know that we get what we measure. If we focus on efficiency measures, we will not produce innovation!

We cannot control our way to innovation. Yet giving up control is not about chaos or allowing a lack of discipline. It is about empowerment and autonomy—a different kind of self-discipline and self-motivation. There is no blueprint for innovation; our mission and goals provide the road map for our direction. And our values guide us in our actions.

We own our failures, learn from them, and then quickly move on. The irony, however, is that to learn, we must often first unlearn. We need to forget what we thought we knew. Though we must be smart about experimentation and leverage our subject matter knowledge, we must often think like a beginner to avoid being blinded by what we think we know.

We need to take an external focus to stay inquisitive. We balance asking questions and staying current on external market trends while being steered internally by the organizations' mission and vision. Part of our external focus includes listening to our customers and regularly soliciting feedback, thereby revising the traditional flow of communication *to* our customers.

The argument has been made that innovation is expensive, but the alternative is significantly more expensive. Tom Kelly of IDEO warned that "in the long run, innovation is cheap. Mediocrity is expensive."[8] The reality, however, is that failure to innovate may result in the biggest cost of all—demise of the organization.

A flatter organizational structure is often seen in more innovative organizations. As organizations grow, however, the bureaucracy and hierarchy of traditional organizational structures is often adopted, often breeding efficiency and standardization, both of which focus on routine processes that squeeze out innovation.

> ### Spotlight on Innovation
>
> "Quality collaboration, in which people have the chance to generate new ideas and work with integrated teams, are a vital part of being innovative, but when the demand to collaborate isn't backed up by quality goals, people's time is being wasted."[9]

We may picture lone individuals toiling away in the dark, but more often, innovation has been proven to be the result of a collaborative effort—being more of a team event.

We may have considered our research and development (R&D) departments as the centers of innovation, but this is no longer the reality. Innovation is everyone's job—across all organizational functions and positions. Letting top management hand the reins of innovation over to a select few fails to communicate the importance of innovation throughout the organization.

We certainly need to address our "old" problems and challenges. However, we need to examine them through a new lens to rethink and reframe them. We cannot continue to view our challenges through the same lens, or we will devise the same solutions—not innovative solutions.

A final paradox is to guard against the "new" ways becoming fossilized or otherwise institutionalized in a rigid way that prohibits our organizations from constantly innovating and moving forward. Our work of innovation is really never done. It is an endless process, one which must be constantly monitored to check our progress.

Culture Assessment: Taking Our Innovative Culture Temperature

We may want to assess our culture broadly to determine the presence of critical elements. Fostering a culture of innovation requires constant renewal. It is not a one-time event, but rather an ongoing process. As a result, we may want to conduct periodic culture audits specifically focused on innovation to see how we are doing. We might consider these questions to take the culture temperature of our organizations:

- Is leadership clearly and continuously communicating the focus on innovation?
- Is leadership walking the talk in modeling behaviors important to innovation?
- Does management remove the organizational barriers that may hinder innovation?
- Is innovation on the formal agenda for most regular meetings?
- Are we specifically hiring creative individuals who fit our innovative culture?
- Do we cultivate an environment where people are comfortable taking risks?
- Do we provide training and development in creative thinking skills?
- Do we reward and recognize innovation?
- Does our performance appraisal specifically address innovation?
- Do we cultivate an environment of playfulness and fun?
- Do we provide workspaces that encourage innovative thinking?
- Do we ourselves innovate? Do we consider ourselves and our organizations as innovative?
- Do we build in autonomy and remove unnecessary control where possible?
- Is HR priming the pump of innovation, or is HR hampering innovation?

And as a final "read," we may want to ask ourselves these questions:

- Am I doing my part to contribute to fostering a culture of innovation?
- Am I an innovator?

A Look in the Mirror: Are We Priming the Pump of Innovation?

As HR professionals, we must consider our own creativity and understand how this creativity might be blocked. A search for the one right answer will stifle creativity. We must be open to a wide variety of alternatives and the recognition that there are usually multiple ways of achieving the results we seek.

We defeat ourselves if we always attempt to be structured, rational, and logical. We are often better served by "winging it" or letting go. Trying to be too controlled in our approach to problem-solving can restrict the flow of those creative juices as we lock into the old ways of doing things. If we stay too regimented, we miss the opportunity to have fun—and to be more creative.

But perhaps the greatest block to creativity is our own self-limiting beliefs. This is essentially the belief that we are not creative and cannot be innovative. In essence, we fail to push boundaries.

Are We Pushing Boundaries?

Innovative individuals (and innovative organizations) are always seeking to push the boundaries. They are energized at the thought of going where they have not been before. Innovators are not shackled to the old ways of doing things. They are flexible and open-minded. However, there must be some reins on their creativity. They want to be flexible but grounded at the same time. They must be creative without losing their sense of reality.

There are innovators who blazed trails across all walks of life. These are the individuals who pushed boundaries. Consider these trailblazers:

- Christopher Columbus (crossing unknown waters).
- Lewis and Clark (traversing uncharted lands).
- Wilbur and Orville Wright (blazing new territory in air transportation).

- Herb Kelleher (who redefined the way we think about flying the "traditional" airline models).
- Mark Zuckerberg (who redefined "friends" and the idea of social media).
- Jeff Bezos (who redefined how and where we shop).
- Peggy Eddens (who entered an all-male club through the kitchen to attend a business luncheon).
- Sandra Day O'Connor (the first woman to sit on the Supreme Court of the United States).
- Fred Smith (who revolutionized the way we ship goods).
- The Beatles (need we say more?).

These are individuals who lived by the motto "nothing ventured, nothing gained." They certainly pushed boundaries. And yet they were grounded in reality—contributing to their success in innovating.

As HR professionals, we certainly need to work within specific parameters such as legal requirements to keep us grounded in reality. Yet we can still innovate within those contexts.

> *Spotlight on Innovation*
> "HR can become the valued enabler and advocate of innovation."[10]

Are We Open-Minded and Comfortable with Change?

Today's environment requires that all of us be more flexible and comfortable with change. The very nature of our world is dynamic and constantly changing. Those who are wed to the past and want things to remain the same quickly find themselves falling behind as the world around them changes.

> *Spotlight on Innovation*
> "The issue of culture change is becoming a bigger part of the HR professional's job."[11]

Those who are fearless are more comfortable with ambiguity. This receptive mindset enables them to take more risks without knowing things with certainty. Decisions are made in three environments: certainty, risk, and uncertainty. Environments of certainty require perfect information. Yet seldom do people have perfect information when making decisions. Even the decision of which car to purchase is usually not made in an environment of complete certainty. Having done our homework, we make our decision only to have a friend ask, "Why didn't you buy a ____?" And of course, we never considered that car (or researched it).

Some people fail to make decisions (or procrastinate) while seeking that information. In academia, students who are in an endless quest for the most complete, up-to-date information for their literature reviews in the dissertation process (constantly adding the latest information) usually end up with the designation ABD—all but dissertation—never finishing their programs. These are the individuals who are not comfortable with ambiguity. Paralysis by analysis closes the door to action.

Spotlight on Innovation

"Creativity is allowing yourself to make mistakes. Art is knowing which ones to keep."[12]

In a rapidly changing, dynamic world it behooves us to become more tolerant of ambiguity. Ironically enough, as formal education prepares students in the "correct" answer, we may not be as open to ambiguous situations and the possibility that there is more than one right answer. Innovative individuals are noted to have a high tolerance for ambiguity.

Today's world calls for innovation and creative problem-solving. Even the paradigm of education has shifted in the last few decades. Once professors were guided by the idea of viewing students' minds as empty vials to be filled with facts and knowledge. That no longer works. Today, professors must teach students how to think. The problems around us are constantly changing, so providing the students with all the answers will serve no useful purpose because the questions and problems will change. The most valuable approach, then, is to teach students how to think (versus what to think). And a large part of teaching students how to think is to focus on creative problem-solving. With more ambiguity and uncertainty, the rational problem-solving approach does not fit most situations. Creative approaches are needed. This, though, requires that people be more fearless as they move out of their comfort zones and chart new territory. To lead the way in fostering an innovative culture, we must move out of our comfort zones.

Are We Taking Risks?

Moving out of one's comfort zone requires a willingness to take risks. Oftentimes, we know what we are good at and do not want to move beyond that for fear that we will not be as good at something else (or heaven forbid, fail at it). This fear of failure holds us back from achieving our full potential. It prevents us from knowing what those boundaries are and what "could have been." We really do not know what we can do until we have tried.

3M has carefully nurtured an innovative organizational culture. Although this has paid off on the bottom line of the corporation, it has not happened by accident. This focus on innovation requires a culture that accepts (and certainly does not punish) failure. If a risk is taken within the organization and it fails, punishment will reinforce that individuals no longer want to take any risks.

The same condition can result from our own self-imposed punishment. If we take a risk and then punish ourselves for failing (by belittling ourselves or hammering our self-esteem), we will no longer take those risks. Robert F. Kennedy once said that "only those who dare to fail greatly can ever achieve greatly."[13] To be fearless, we must give ourselves permission to fail!

Are We Conquering Fear?

We cannot let fear paralyze us. It is important to take action and to take that first step. Bill Meredith, a career military man, once told his children that they could fear *one* thing—and only one thing. Hearing his children use the excuse "I'm afraid" over and over again made him come to the conclusion that fear held people back in a number of areas. Children could not excel in school, in athletics, or in life if they were allowed to use their fear as an excuse. Though picking one fear may seem a little extreme, it certainly focused his children on learning the importance of not letting fear hold them back—and of facing those fears.

Sometimes we just need to ask what we really fear. What are we truly afraid of? Articulating this fear can make it more manageable and therefore easier to overcome. Dale Carnegie said, "Inaction breeds doubt and fear. Action breeds confidence and courage. If you want to conquer fear, do not sit at home and think about it. Go out and get busy."[14] Carnegie further recommended that we go out and keep doing the thing we fear! It is human to have fears. Letting them multiply and drive our life, though, is self-limiting. The difference between successful people and those who are not successful is in the response to their fears. Successful people face their fears—yes, they have them! Those who are not successful let their fears hold them back.

Fear can be perceived or real. Either is equally debilitating. Subconscious fear is just as powerful in eroding our self-confidence as real fear. Furthermore, this fear prohibits us from taking the first step to realize our dreams. When people are asked what their greatest fears are, it is always amazing to see results that indicate a fear of public speaking is ranked higher by most people than a fear of dying! Perhaps this gives credence to the idea that somehow many of our fears are unreasonable and not grounded in reality.

We can see, then, why fear is the antidote to the innovation we seek. Fear holds us back from taking risks, from experimenting—and ultimately, from innovating. Fear drives out innovation. Fear of evaluation closes down the pipeline of new ideas and opens us up to killing ideas with skepticism. With fear, there is no supportive environment to nurture our ideas and receive constructive feedback. Fear of giving up control crushes the spirit of those who are innovators and crave that freedom.

As HR professionals, we need to expect to do the heavy lifting. We must consider taking that all-important first step. We cannot wait for a critical mass to get behind our effort. We may need to be that lone voice at the front leading the way. By taking the temperature of our organizational cultures and assessing our own readiness to stimulate innovation, we are better positioned to lead the charge in accepting the challenge to foster an innovative culture!

Spotlight on Innovation

"Only individuals and teams innovate. Organizations, cultures, and processes do not innovate; they can only support or inhibit the individuals and teams who want to innovate."[15]

Endnotes

Chapter 1

1. Amy C. Edmondson, *Teaming to Innovate* (San Francisco: Jossey-Bass, 2013), 28.
2. Gary Hamel and C. K. Prahalad, *Competing for the Future* (Boston: Harvard Business Review Press, 1994).
3. Matt Donovan, "Shifting Focus to Agile Development," *Talent Management*, November 13, 2014, 30-33, http://www.talentmgt.com/articles/6943-shifting-focus-to-agile-development.
4. Ibid.
5. Michael Stanleigh, "Innovation: A Strategic HR Imperative," Business Improvement Architects, 2012, http://www.bia.ca/articles/InnovationAStrategicHRImperative.htm.
6. ASTD Research and Claude Legrand, *Building an Innovative Organization: The Role of Training and Development*, 2014, https://www.td.org/Publications/Research-Reports/2014/Building-An-Innovative-Organization.
7. Stanleigh, "Innovation."
8. "The Most Innovative Companies," BCG Perspectives, October 28, 2014, https://www.bcgperspectives.com/content/interactive/innovation_growth_most_innovative_companies_interactive_guide.
9. Ernest Gundling, *The 3M Way to Innovation: Balancing People and Profit* (New York: Kodansha International, 2000), 23.
10. Patty Gaul, "Innovation Matters, but Prioritizing It Lags," *TD Magazine*, April 8, 2014, 22, https://www.td.org/Publications/Magazines/TD/TD-Archive/2014/04/Innovation-Matters-but-Prioritizing-It-Lags.
11. David Burkus, "10 Practices That Drive Innovation," LDRLB, April 10, 2013, http://ldrlb.co/2013/04/10-practices-that-drive-innovation.

12. Stanleigh, "Innovation."

13. Alex Gammelgard, "Three Ideas for Encouraging Workplace Innovation," Arena Solutions, http://www.arenasolutions.com/blog/post/workplace-innovation.

14. Rosabeth Moss Kanter, "Innovation: The Classic Traps," *Harvard Business Review*, November 2006, 73-83, https://hbr.org/2006/11/innovation-the-classic-traps.

15. Stanleigh, "Innovation."

16. Gammelgard, "Three Ideas for Encouraging Workplace Innovation."

17. Paul Nunes and Tim Breene, "Reinvent Your Business Before It's Too Late," *Harvard Business Review*, January 2011, 82, https://hbr.org/2011/01/reinvent-your-business-before-its-too-late.

18. Stanleigh, "Innovation."

19. Tony Hsieh, "Culture is Priority One," YouTube, June 12, 2011, https://www.youtube.com/watch?v=-4D3RplqmyU.

20. Ross Tartell, "If 'Culture' is Key, How Can Training Help?," *Training*, July/August 2014, 10-11, http://www.trainingmag.com/trgmag-article/if-%E2%80%9Cculture%E2%80%9D-key-how-can-training-help. Also, noted management guru Peter Drucker has been credited for the well-known quote: "culture eats strategy for lunch."

21. Neil Anderson, Kristina Potočnik, and Jing Zhou, "Innovation and Creativity in Organizations: A State-of-the-Science Review, Prospective Commentary, and Guiding Framework," *Journal of Management* 40(5), July 2014, 1298.

22. Stanleigh, "Innovation."

Chapter 2

1. Scott Edinger, "Don't Innovate. Create a Culture of Innovation," *Forbes*, November 20, 2012, http://www.forbes.com/sites/scottedinger/2012/11/20/dont-innovate-create-a-culture-of-innovation.

2. Neil Anderson, Kristina Potočnik, and Jing Zhou, "Innovation and Creativity in Organizations: A State-of-the-Science Review,

Prospective Commentary, and Guiding Framework," *Journal of Management* 40(5), July 2014, 1298.

3. Robert Hartland, "How to Make Innovative Ideas Happen," *Smashing Magazine*, October 22, 2010, http://www.smashing-magazine.com/2010/10/22/how-to-make-innovative-ideas-happen.

4. Ernest Gundling, *The 3M Way to Innovation: Balancing People and Profit* (New York: Kodansha International, 2000), 58.

5. Daniel Isenberg, "Entrepreneurs and the Cult of Failure," *Harvard Business Review*, April 2011, 36, https://hbr.org/2011/04/column-entrepreneurs-and-the-cult-of-failure.

6. Amy C. Edmondson, "Strategies for Learning from Failure," *Harvard Business Review*, April 2011, 47-55, https://hbr.org/2011/04/strategies-for-learning-from-failure.

7. Isenberg, "Entrepreneurs and the Cult of Failure."

8. Edmondson, "Strategies for Learning from Failure," 55.

9. Rita Gunther McGrath, "Failing by Design," *Harvard Business Review*, April 2011, 78, https://hbr.org/2011/04/failing-by-design.

10. Karen Dillon, "I Think of My Failure as a Gift," *Harvard Business Review*, April 2011, 86-9, https://hbr.org/2011/04/i-think-of-my-failures-as-a-gift.

11. Francesca Gino and Gary P. Pisano, "Why Leaders Don't Learn from Success," *Harvard Business Review*, April 2011, 68-74, https://hbr.org/2011/04/why-leaders-dont-learn-from-success.

12. Ibid., 71.

13. Ibid., 73.

14. Ibid., 74.

15. Robert Todd, Lisa Buckley, and Sam Herring, "The Innovator Chief Learning Officer," *TD Magazine*, November 8, 2014, 36, https://www.td.org/Publications/Magazines/TD/TD-Archive/2014/11/The-Innovator-Chief-Learning-Officer.

16. Catherine H. Tinsley, Robin L. Dillon, and Peter M. Madsen, "How to Avoid Catastrophe," *Harvard Business Review*, April 2011, 96, https://hbr.org/2011/04/how to avoid-catastrophe.

17. Gundling, *The 3M Way to Innovation*, 187.

18. Rosabeth Moss Kanter, "Innovation: The Classic Traps," *Harvard Business Review*, November 2006, 80, https://hbr.org/2006/11/innovation-the-classic-traps.

19. Osvald M. Bjelland and Robert Chapman Wood, "An Inside View of IBM's 'Innovation Jam'," *MIT Sloan Management Review*, Fall 2008, 32-40, http://sloanreview.mit.edu/article/an-inside-view-of-ibms-innovation-jam.

20. Amy C. Edmondson, *Teaming to Innovate* (San Francisco: Jossey-Bass, 2013), 62-63.

21. "A Century of Innovation: The 3M Story," 2002, 21, http://multimedia.3m.com/mws/media/171240O/3m-coi-book-tif.pdf?fn=3M_COI_Book.pdf.

22. Daniel H. Pink, *Drive: The Surprising Truth about What Motivates Us* (New York: Riverhead Books, 2009).

23. Sarah Fister Gale, "Collaboration Realization," *Workforce*, December 3, 2014, 45, http://www.workforce.com/articles/20959-collaboration-realization.

24. Kanter, "Innovation," 76.

25. Anderson, Potočnik, and Zhou, "Innovation and Creativity in Organizations," 1321.

26. Rosabeth Moss Kanter, "Nine Rules for Stifling Innovation," *Harvard Business Review*, January 15, 2013, https://hbr.org/2013/01/nine-rules-for-stifling-innova.

Chapter 3

1. David Burkus, "10 Practices That Drive Innovation," LDRLB, April 10, 2013, http://ldrlb.co/2013/04/10-practices-that-drive-innovation.

2. Rosabeth Moss Kanter, "Innovation: The Classic Traps," *Harvard Business Review*, November 2006, 82, https://hbr.org/2006/11/innovation-the-classic-traps.

3. Luke Siuty, "Designing a Diverse Culture," *Diversity Executive*, November 13, 2014, http://www.talentmgt.com/articles/6944-designing-a-diverse-culture.

4. "Zappos Family Core Values," Zappos, http://about.zappos.com/our-unique-culture/zappos-core-values; Holacracy can be found at www.holacracy.org.

5. Terry M. Farmer and Xavier Butte, "Inspire to Innovate," *TD Magazine*, September 8, 2014, 57, https://www.td.org/Publications/Magazines/TD/TD-Archive/2014/09/Inspire-to-Innovate.

6. Luke Siuty, "Designing a Diverse Culture."

7. Amy C. Edmondson, *Teaming to Innovate* (San Francisco: Jossey-Bass, 2013).

8. Griffith and Dunham, *Working in Teams*, 140.

9. Rich Horwath, Deep Dive: The Proven Method for Building Strategy, Focusing Your Resources, and Taking Smart Action, 2nd edition (Austin, TX: Greenleaf, 2009).

10. Edmondson, *Teaming to Innovate*, 17.

11. William Sebra, "Innovation Gives HR a Leg Up," *Talent Management*, March 28, 2014, 44-47, http://www.talentmgt.com/articles/innovation-gives-hr-a-leg-up.

12. Andrew R. McIlvaine, "Powering a Revolution," *Human Resource Executive*, July 21, 2014, 14-7, http://www.hreonline.com/HRE/view/story.jhtml?id=534357301.

13. Michael Stanleigh, "Innovation: A Strategic HR Imperative," Business Improvement Architects, 2012, http://www.bia.ca/articles/InnovationAStrategicHRImperative.htm.

14. For example, see Tovia Smith, "Does Teaching Kids To Get 'Gritty' Help Them Get Ahead?," NPR, March 17, 2014, http://www.npr.org/blogs/ed/2014/03/17/290089998/does-teaching-kids-to-get-gritty-help-them-get-ahead.

15. Robert Todd, Lisa Buckley, and Sam Herring, "The Innovator Chief Learning Officer," *TD Magazine*, November 8, 2014, 36, https://www.td.org/Publications/Magazines/TD/TD-Archive/2014/11/The-Innovator-Chief-Learning-Officer, 41.

16. Griffith and Dunham, *Working in Teams*, 140.

17. This quote is attributed to Albert Einstein.

18. Sebra, "Innovation Gives HR a Leg Up."

19. Ibid; also see "The Innovation Imperative: A Global Insight Study Investigating the Impact of Innovation on Recruitment and Talent Management," Futurestep, June 2013, http://www.futurestep.com/opinions/the-innovation-imperative.

Chapter 4

1. ASTD Research and Claude Legrand, *Building an Innovative Organization: The Role of Training and Development*, 2014, https://www.td.org/Publications/Research-Reports/2014/Building-An-Innovative-Organization.

2. Scott Edinger, "Don't Innovate. Create a Culture of Innovation," *Forbes*, November 20, 2012, http://www.forbes.com/sites/scottedinger/2012/11/20/dont-innovate-create-a-culture-of-innovation.

3. Robert Todd, Lisa Buckley, and Sam Herring, "The Innovator Chief Learning Officer," *TD Magazine*, November 8, 2014, 36-41, https://www.td.org/Publications/Magazines/TD/TD-Archive/2014/11/The-Innovator-Chief-Learning-Officer.

4. Neil Anderson, Kristina Potočnik, and Jing Zhou, "Innovation and Creativity in Organizations: A State-of-the-Science Review, Prospective Commentary, and Guiding Framework," *Journal of Management* 40(5), July 2014, 1297.

5. Daniel Denison, Ia Ko, Lindsey Kotrba, and Levi Nieminen, "Drive an Innovative Culture," *Chief Learning Officer*, June 3, 2013, 70-72, http://www.clomedia.com/articles/drive-an-innovative-culture.

6. Fionna Patterson, Máire Kerrin, and Geraldine Gatto-Roissard, *Characteristics and Behaviours of Innovative People in Organisations: Literature Review*, NESTA, 2, http://www.nesta.org.uk/sites/default/files/characteristics_behaviours_of_innovative_people.pdf.

7. John Boudreau, "Work Deconstructed," *Talent Management*, June 3, 2014, 10, http://www.talentmgt.com/articles/4732.

8. Gary Hamel and C. K. Prahalad, *Competing for the Future* (Boston: Harvard Business Review Press, 1994); Scott Elser, "6 Words Your

Employees Say That Will Kill Your Business," Inc., July 1, 2014, http://www.inc.com/scott-elser/6-words-your-employees-say-that-will-kill-your-business.html.

9. John J. Hampton, Fundamentals of Enterprise Risk Management: How Top Companies Assess Risk, Manage Exposure, and Seize Opportunity, 2nd edition (New York: AMACOM, 2015), p. 196.

10. Daniel H. Pink, *A Whole New Mind: Why Right-Brainers Will Rule the Future* (New York: Riverhead Books, 2006).

11. Tina Su, "7 Habits of Highly Innovative People," Think Simple Now, http://thinksimplenow.com/creativity/7-habits-of-highly-innovative-people.

12. Grace McGartland, "12 Ways to be More Creative at Work," Thunderbolt Thinking, http://www.thunderboltthinking.com/12ways.htm.

13. Bruna Martinuzzi, "Optimism: The Hidden Asset," Mind Tools, http://www.mindtools.com/pages/article/newLDR_72.htm.

14. Alan Loy McGinnis, "12 Characteristics of Tough-Minded Individuals," in *The Power of Optimism* (1993), with commentary by Henry Givray, 2011, http://leadersforum2011.smithbucklin.com/Portals/4/content/2011/12_Characteristics_of_Tough_Minded_Optimists.pdf.

15. The quote is attributed to Winston Churchill.

16. Su, "7 Habits of Highly Innovative People."

17. Marcus Buckingham, *The Truth About You: Your Secret to Success* (Nashville, TN: Thomas Nelson, 2008).

18. Todd, Buckley, and Herring, "The Innovator Chief Learning Officer," 39.

19. Ibid.

20. For assistance, see Tony Buzan with Barry Buzan, The Mind Map Book: How to Use Radiant Thinking to Maximize Your Brain's Untapped Potential (NY: Plume, 1996).

21. Rosabeth Moss Kanter, "Innovation: The Classic Traps," *Harvard Business Review*, November 2006, 82, https://hbr.org/2006/11/innovation-the-classic-traps.

Chapter 5

1. "Chapter 15: Driving a Culture of Innovation," The Build Network, October 11, 2013, http://thebuildnetwork.com/leadership/innovation-culture.
2. Systematic Inventive Thinking, *How Companies Incentivize Innovation*, April 2013, 1, http://www.innovationinpractice. com/How%20Companies%20Incentivize%20Innovation%20 E-version%20May%202013.pdf.
3. Ann Bares, "Innovate or Perish: The Role of Rewards," Compensation Café, July 15, 2011, http://www.Compensationcafe. com/2011/07/innovate-or-perish-the-role-of-rewards.html.
4. Paige Leavitt, "Rewarding Innovation," American Productivity Quality Center, http://www.providersedge.com/docs/km_articles/rewarding_innovation.pdf.
5. Systematic Inventive Thinking, "How Companies Incentivize Innovation," April 2013, 1, http://www.innovationinpractice. com/How%20Companies%20Incentivize%20Innovation%20 E-version%20May%202013.pdf.
6. Teresa M. Amabile, "How to Kill Creativity," *Harvard Business Review*, September 1998, 76–87, https://hbr.org/1998/09/how-to-kill-creativity/ar/1.
7. For example, see Bianca Lavorta, "Undermining Intrinsic Motivation," *SFU Ed Review*, November 1, 2013, http://www.sfuedreview.org/undermining-intrinsic-motivation.
8. Alex Gamelgaard, "Three Ideas for Encouraging Workplace Innovation," Arena Solutions, http://www.arenasolutions.com/blog/post/workplace-innovation.
9. Leavitt, "Rewarding Innovation," 2.
10. Drew Gannon, "How to Reward Great Ideas," *Inc.*, July 19, 2011, http://www.inc.com/guides/201107/how-to-reward-employees-great-ideas.html; also see Maritz Research, "Managing in an Era of Mistrust," Maritz Poll, news release, April 14, 2010, http:// www.maritzresearch.com/shared-content/Maritz-Poll/2010/ Maritz-Poll-Reveals-Employees-Lack-Trust-in-their-Workplace.
11. Bares, "Innovate or Perish."

12. Tomislav Buijubasic, "Rewarding Innovation," Innovation Excellence, August 4, 2013, http://www.innovationexcellence.com/blog/2013/08/04/rewarding-innovation-2/. Emphasis in original.

13. Michael Stanleigh, "Innovation: A Strategic HR Imperative," Business Improvement Architects, http://www.bia.ca/articles/InnovationAStrategicHRImperative.htm.

14. Systematic Inventive Thinking, *How Companies Incentivize Innovation*, 12.

15. By Seth Waugh, CEO of Deutsche Bank America as quoted in Stanleigh, "Innovation."

Chapter 6

1. Gary Hamel, "The Why, What, and How of Management Innovation," *Harvard Business Review*, February 2006, 72-84, https://hbr.org/2006/02/the-why-what-and-how-of-management-innovation.

2. Soren Kaplan, "6 Ways to Create a Culture of Innovation," *Fast Company*, December 21, 2013, http://www.fastcodesign.com/1672718/6-ways-to-create-a-culture-of-innovation.

3. Amy C. Edmondson, *Teaming to Innovate* (San Francisco: Jossey-Bass, 2013), 43.

4. Ibid., 104.

5 Ibid., 91.

6. Keith Naughton, "The Happiest Man in Detroit," Bloomberg Business, February 3, 2011, http://www.bloomberg.com/bw/magazine/content/11_07/b4215066125842.htm; also Edmondson, *Teaming to Innovate*, 105.

7. A.G. Lafley and Ram Charan, *The Game-Changer: How You Can Drive Revenue and Profit Growth with Innovation* (New York: Random House, 2008), 210.

8. Edmondson, *Teaming to Innovate*, 64.

9. Terry M. Farmer and Xavier Butte, "Inspire to Innovate," *TD Magazine*, September 8, 2014, 59, https://www.td.org/Publications/Magazines/TD/TD-Archive/2014/09/Inspire-to-Innovate.

10. Ibid., 58.

11. Gary Hamel, *Matters Now: How to Win in a World of Relentless Change, Ferocious Competition, and Unstoppable Innovation* (San Francisco: Jossey-Bass, 2012).

12. Daniel H. Pink, *Drive: The Surprising Truth about What Motivates Us* (New York: Riverhead Books, 2011).

13. Scott Edinger, "Don't Innovate. Create a Culture of Innovation," *Forbes*, November 20, 2012, http://www.forbes.com/sites/scottedinger/2012/11/20/dont-innovate-create-a-culture-of-innovation.

14. Thomas J. Peters and Robert H. Waterman, *In Search of Excellence: Lessons from America's Best-Run Companies* (New York: Harper & Row, 1982).

15. Pink, *Drive*, 105.

16. Ibid., 95.

17. "How Do You Build a Culture of Innovation?" A video interview with Tim Brown, Yale Insights, http://insights.som.yale.edu/insights/how-do-you-build-culture-innovation.

18. Max Mihelich, "Goin' ROWE at Edmunds.com," *Talent Management*, May 27, 2014, 36-41, http://www.talentmgt.com/articles/goin-rowe.

19. Pink, *Drive*, 93.

20. David P. Sorensen, *Innovations: Keys to Success* (Menlo Park, CA: Crisp Publications, 1997).

21. Rob Cross and Chris Ernst, "Deploying Network Talent to Drive Innovation," *Workforce*, August 29, 2014, 26, http://www.workforce.com/articles/20755-deploying-network-talent-to-drive-innovation.

22. Rita Pyrillis, "Searching for Solace," *Workforce*, November 5, 2014, 41, http://www.workforce.com/articles/20868-searching-for-solace.

23. Harvey Meyer, "Future Digs," Human Resource Executive Online, September 30, 2014, http://www.hreonline.com/HRE/view/story.jhtml?id=534357647.

24. Luke Siuty, "Managers' Style Stifles Culture Change," *Talent Management*, December 12, 2014, 44, http://www.talentmgt.

com/articles/6994-managers-style-stifles-culture-change. Note that Siuty is reporting on the findings in a survey from Human Capital Media Advisory Group.

25. Some companies have used a free application called Evernote, which can be found at https://evernote.com.

26. Tomas Chamorro-Premuzic, "Why Brainstorming Works Better Online," Harvard Business Review, April 2, 2015, https://hbr. org/2015/04/why-brainstorming-works-better-online.

27. Farmer and Butte, "Inspire to Innovate," 54.

28. Cross and Ernst, "Deploying Network Talent to Drive Innovation," 26.

29. Brian Griffith and Ethan Dunham, *Working in Teams: Moving from High Potential to High Performance* (Thousand Oaks, CA: Sage, 2015), 142.

30. John Sullivan, "Why Most Firms Have Given Up on Matching the Talent Features at Google," TLNT, March 19, 2014, http://www. tlnt.com/2014/03/19/why-most-firms-have-given-up-on-match-ing-the-talent-features-at-google.

Chapter 7

1. Keith H. Hammonds, "Why We Hate HR," *Fast Company*, August 2005, http://www.fastcompany.com/53319/why-we-hate-hr.

2. Tom A. Sims, "Towards a Customer Centric Talent Strategy," *Talent Management Excellence Essentials*, June 2014, 18, http://www.hr.com/en/topleaders/interactive_content/talent-management-excellence-essentials-june-2014_hwn8of4z.html.

3. Linda David, "Key HR Initiatives From 2014," *TD Magazine*, October 8, 2014, https://www.td.org/Publications/Magazines/TD/TD-Archive/2014/10/Intelligence-Key-HR-Initiatives-from-2014; the full report can be found at Towers Watson, *2014 Global Talent Management and Rewards Study—At a Glance*, 2014, http://www.towerswatson.com/DownloadMedia.aspx?media={B2A765FF-2895-450B-A636-EF1F3837ACA1}.

4. William Sebra, "Innovation Gives HR a Leg Up," *Talent Management*, March 28, 2014, 44-47, http://www.talentmgt.com/articles/innovation-gives-hr-a-leg-up.

5. Jac Fitz-enz, "HR Steps Up Its Game," *Talent Management*, October 10, 2014, 12, http://www.talentmgt.com/articles/6867-hr-steps-up-its-game.

6. John Boudreau and Ed Lawler, "Is HR Ready to Face the Future?" *Talent Management*, June 5, 2014, 47, http://www.talentmgt.com/articles/is-hr-ready-to-face-the-future-1.

7. Leslie Caccamese, "The 8 Cultural Characteristics that Make Google a Great Innovator," Great Place to Work, April 26, 2012, http://www.greatplacetowork.com/publications-and-events/blogs-and-news/978.

8. Andrew R. McIlvaine, "Powering a Revolution," *Human Resource Executive*, July 21, 2014, 14, http://www.hreonline.com/HRE/view/story.jhtml?id=534357301.

9. Michael Stanleigh, "Innovation: A Strategic HR Imperative," Business Improvement Architects, 2012, http://www.bia.ca/articles/InnovationAStrategicHRImperative.htm.

10. Jac Fitz-enz, "What It Really Means to Be Engaged," *Talent Management*, November 5, 2014, 12, http://www.talentmgt.com/articles/6926-what-it-really-means-to-be-engaged.

11. Sharon Jordan-Evans and Beverly Kaye, "More Stay Interviews, Fewer Exit Interviews," *Talent Management*, September 23, 2014, 20-23, http://www.talentmgt.com/articles/6828-more-stay-interviews-fewer-exit-interviews; also see Richard P. Finnegan, *The Power of Stay Interviews for Engagement and Retention* (Alexandria, VA: Society for Human Resource Management, 2012).

12. *The People Side of Great Business with Libby Sartain*, Stanford Executive Briefings, Kantola Productions, 51 min., 2004, dvd.

13. Ashok Som, "Emerging Human Resource Practices at Aditya Birla Group," *Human Resource Management*, 49(3), May-June 2010, 549.

14. For a definition and description of hackathon, see "Hackathon," Wikipedia, en.wikipedia.org/wiki/Hackathon.

15. Som, "Emerging Human Resource Practices," 564-65.

16. Teresa M. Amabile, "How to Kill Creativity," *Harvard Business Review*, September 1998, 77-87, https://hbr.org/1998/09/how-to-kill-creativity/ar/1.

17. Adam Bryant, "Google's Quest to Build a Better Boss," *New York Times*, March 12, 2011, http://www.nytimes.com/2011/03/13/business/13hire.html?_r=1.

18. Robert Todd, Lisa Buckley, and Sam Herring, "The Innovator Chief Learning Officer," *TD Magazine*, November 8, 2014, 39, https://www.td.org/Publications/Magazines/TD/TD-Archive/2014/11/The-Innovator-Chief-Learning-Officer.

19. Stanleigh, "Innovation."

20. Carol Morrison, "The Learning Function's Latest Role: Organizational Change Agent," TD Magazine, October 8, 2014, 24, https://www.td.org/Publications/Magazines/TD/TD-Archive/2014/10/The-Learning-Functions-Latest-Role-Organizational-Change-Agent.

Chapter 8

1. Daniel Denison, Ia Ko, Lindsey Kotrba, and Levi Nieminen, "Drive an Innovative Culture," *Chief Learning Officer*, June 3, 2013, 70-2, http://www.clomedia.com/articles/drive-an-innovative-culture.

2. Ernest Gundling, *The 3M Way to Innovation: Balancing People and Profit* (New York: Kodansha International, 2000), 12.

3. Evodio Kaltenecker, "Innovation Chat with Scott Anthony," Innovation Excellence, March 20, 2013, http://www.innovation-excellence.com/blog/2013/03/20/innovation-chat-with-scott-anthony.

4. See Clayton M. Christensen, The Innovator's Dilemma: When New Technologies Cause Great Firms To Fail (Boston: Harvard Business School Press, 1997); and Clayton M. Christensen and Michael E. Radnor, The Innovator's Solution: Creating and Sustaining Successful Growth (Boston: Harvard Business School Press, 2003).

5. Michael Stanleigh, "Innovation: A Strategic HR Imperative," Business Improvement Architects, 2012, http://www.bia.ca/articles/InnovationAStrategicHRImperative.htm.

6. Rosabeth Moss Kanter, "Innovation: The Classic Traps," *Harvard Business Review*, November 2006, 80, https://hbr.org/2006/11/innovation-the-classic-traps.

7. Jim Collins and Jerry Porras, *Built to Last: Successful Habits of Visionary Companies* (New York: HarperBusiness, 2004) 43-44.

8. Daniel H. Pink, *Drive: The Surprising Truth about What Motivates Us* (New York: Riverhead Books: 2009) 88.

9. Sarah Fister Gale, "Collaboration Realization," *Workforce*, December 3, 2014, 49, http://www.workforce.com/articles/20959-collaboration-realization.

10. Michael J. Wolf, "How HR Departments Can Create an Innovation Culture," *Wall Street Journal*, May 2, 2014, http://blogs.wsj.com/experts/2014/05/02/how-hr-departments-can-create-an-innovation-culture.

11. Luke Siuty, "Managers' Style Stifles Culture Change," *Talent Management*, December 12, 2014, 44, http://www.talentmgt.com/articles/6994-managers-style-stifles-culture-change.

12. The quote is attributed to cartoonist Scott Adams.

13. "Primary Resources: RFK in Capetown," *American Experience*, PBS, http://www.pbs.org/wgbh/americanexperience/features/primary-resources/kennedys-capetown.

14. "20 Tips on Overcoming Fear," Dale Carnegie Training, June 5, 2009, http://blog.dalecarnegie.com/tipsforsuccess/20-tips-on-overcoming-fear.

15. ASTD Research and Claude Legrand, *Building an Innovative Organization: The Role of Training and Development*, 2014, https://www.td.org/Publications/Research-Reports/2014/Building-An-Innovative-Organization.

Index

?What If!, 49
"10 Practices That Drive Innovation,"
    33
The 3M Way to Innovation, 5
3M, 5, 10, 14, 15, 19, 23, 27, 31, 63,
    64, 67, 81, 84, 95, 99, 113
7 Habits of Highly Innovative People, 56

A

adaptability, 33, 50
adaptable, 28, 43, 44, 74, 78
after-action reviews, 21
agility, 2, 92
Altura Consulting Group, 62
Amabile, Teresa M., 96
Amazon.com, 3
American Greetings, 95
American Productivity and Quality
    Center (APQC), 62
Anthony, Scott, 102
Apple, 2, 3, 7, 98, 103
Apttus, 8, 64
Association for Talent Development,
    90
ASTD, 3, 49
Atlassian, 79
autonomy, 27, 28, 31, 34, 35, 44, 60,
    64, 79, 81, 94, 107, 109

B

BankBoston, 28
Bares, Ann, 62
Beatles, The, 111
behavior(s), 18, 26, 49, 56, 61, 62,
    63, 71, 74, 77, 78, 79, 85, 87, 88,
    89, 92, 93, 99, 101, 105, 109
Bell, Alexander Graham, 57
Berkun, Scott, 54
Best Buy, 80, 103

Bezos, Jeff, 111
Bjelland, Osvald M., 26
BMW, 6
Boston Consulting Group (BCG), 3
Boston Red Sox, 94
Boudreau, John, 91
Bowerman, Bill, 98
brainstorm, 46, 54, 65
brainstorming, 26, 83, 85, 95, 96
    sessions, 59, 85
    skills, 59
    techniques, 58
branding, 34
Breene, Tim, 9
Buckingham, Marcus, 56
Buckley, Lisa, 49
budgeting, 23, 28
Building an Innovative Organization, 3, 49
Burkus, David, 6, 33
Business Improvement Architects,
    98
business performance, 1
Butte, Xavier, 36, 78, 86

C

career paths, 57
Carnegie, Dale, 114
Carrier, Gordon, 36
Center for Effective Organizations,
    89, 90
change management, 99
chaos, 16, 28, 81, 107
Children's Hospitals and Clinics of
    Minnesota, 17, 75
Chipotle Mexican Grill, 3
Circuit City, 103
Clarion Enterprises, 55
coaching, 59, 87, 97
Coca-Cola, 3
collaborating, 24

collaboration, 6, 25, 27, 31, 33, 40, 46, 60, 64, 73, 81, 82, 83, 95, 105, 108
collaborative
   effort(s), 1, 108
   element, 45
   environment, 13, 24, 25
   inquiry, 58
   nature, 59
   process, 3
   relationships, 27
   spaces, 35
   thinking, 25
Collins, Jim, 106
Columbus, Christopher, 111
communication, 13, 23, 25, 26, 51, 99, 104, 107
   defensive, 56
   downward, 24
   effective, 20
   nonverbal, 23
   open, 24, 25, 26, 30, 78, 105
   persuasive, 47
   process, 84
   skills, 46, 47, 59
   style, 87
   top-down, 83
   transparent, 24, 73
Competing for the Future, 2, 53
competitive advantage, 2, 7, 10, 13, 31, 41, 95, 103
confidence, 19, 20, 37, 114
   over, 20
   self-, 8, 64, 114
constructive feedback, 115
constructive stumbling, 19
courage, 15, 114
creative, 27, 29, 37, 50, 52, 56, 63, 102, 110
   applicants, 92
   approaches, 113
   destruction, 103
   floor plan, 83
   freedom, 35
   ideas, 36
   individuals, 33, 34, 38, 43, 44, 46, 47, 55, 94-5, 106, 109
   insights, 57
   minds, 33

   people, 45
   personality, 26
   problem-solving, 112-13
   process, 25, 38, 45, 46, 59
   rooms, 84
   thinker(s), 52,
   thinking, 4, 52, 96, 106, 109
   types, 34, 45, 59, 96
   zone, 57
creativity, 4, 11, 27, 28, 29, 30, 35, 37, 45, 46, 47, 50, 54, 56, 59, 74, 79, 83, 96, 99, 101, 104, 110, 112
Cross, Rob, 82, 86
cross-functional teams, 25, 28, 31, 82, 87
cultural onboarding, 91
culture(s)
adaptive, 10, 103
assessment, 108
change, 111
company, 63, 95
corporate, 10, 11, 13, 31, 33, 69
culture, 108-09
misfits, 35
organizational, 3, 8, 15, 17, 50, 61, 63, 99, 113, 115
supportive, 33
training, 100
unsupportive, 10
workplace, 84
curiosity, 28, 33, 37, 41, 42, 44, 49, 57, 86

D

deconstruction, 58
development, 11, 49, 50, 54, 58, 59, 73, 92, 96, 99, 109
   leadership, 30, 59, 74
   management, 66
   opportunities, 68
   plans, 57
   professional, 88
Dillon, Karen, 18
Dillon, Robin L., 22
Discovery Health, 44
Disney, 10
   Keys to Excellence, 10, 35
disruption, 103
disruptive, 8

diverse workforce(s), 33, 87
diversity, 36-7, 38, 77, 87, 104
Donovan, Matt, 2
Dorsey, Jack, 58
"Drive an Innovative Culture," 50, 101
Drive, 27
Ducati, 20
Dunham, Ethan, 45

E

Eddens, Peggy, 111
Edinger, Scott, 13, 49
Edmondson, Amy C., 1, 16, 38, 75
effective workspaces, 83
"Eight Habits of Highly Effective Google Managers," 97
EiQ, 36, 78
Eli Lilly, 24, 75
entrepreneurial spirit, 48
Ernst, Chris, 82, 86
experimentation, 43, 57, 59, 105, 107

F

Facebook, 95
failure(s), 13-22, 24, 27, 42, 43, 47, 55, 57, 63, 66, 75, 76, 77, 78, 84, 88, 105, 106, 107, 113
    intelligent, 16, 17, 24, 43, 45, 75
    instructive, 16
Farmer, Terry M., 36, 78, 86
Fast Company, 89
fear(s), 15, 24, 76, 88, 113, 114, 115
First Community Bank, 28-9
Fitz-enz, Jac, 91
flexibility, 27, 29, 30, 67, 87
Forbes, 3
Ford, 76
Foursquare, 70
freedom, 27, 29, 30, 34, 35, 44, 54, 57, 64, 79, 80, 81, 84, 94, 95, 115
Frima, 67
Futurestep, 90

G

Gammelgard, Alex, 8, 64
Gates, Bill, 53

GE Capital Real Estate-North America, 10
General Motors, 29
Gensler, 83
Georgetown University Hospital, 79
Gino, Francesca, 19, 21
GlaxoSmithKline, 83, 103
Google, 3, 28, 34, 65, 97, 106
Griffith, Brian, 45
Gundling, Ernest, 5, 15

H

hackathon(s), 95, 97
Hamel, Gary, 2, 53, 73, 78
Hammonds, Keith, 89
Harris Group, 3
Harvard Business Review, 15, 18
Herring, Sam, 49
hierarchy, 27, 28, 29, 79, 80, 105, 107
hiring, 33, 34, 35, 36, 37, 42, 64, 91, 95, 96, 109
holacracy, 35
Hsieh, Tony, 10, 35
Human Capital Performance Partners, 45
Human Capital Source, 91
human capital, 36, 42, 73
human resources (HR), 6, 8, 11, 12, 33, 40, 43, 48, 49, 89, 90, 91, 92, 94, 95, 96, 97, 98, 100, 109, 111
    department(s), 69, 90, 91, 92, 93-4, 95
    function, 94
    industry, 90
    innovation, 90
    leaders, 11, 43, 91
    manager, 41
    perspective, 40
    practices, 21, 48, 98
    practitioners, 84
    professionals, 11, 48, 56, 89, 90, 91, 92, 96, 97, 98, 99, 101, 102, 110, 111, 115
    unit, 91

I

IBM Global CEO, 10

IBM, 3, 26, 92
Ideaction, 6
IDEO, 14, 54, 84, 106, 107
In Search of Excellence, 79
Innosight, 102
Innovation Imperative Study, 47
interconnectedness, 39, 52, 91
interpersonal skills, 33
Intrepid Learning, 49
Intuit, 23
intuition, 52, 54, 106
iPhone, 7
Isenberg, Daniel, 15, 16

J

Jarrett, Keith, 104
job descriptions, 27, 97, 105
Jobs, Steve, 38
Jordan-Evans, Sharon, 93
Juniper Networks, 82

K

Kanter, Rosabeth Moss, 7, 8, 28, 106
Kaye, Beverly, 93
Kelleher, Herb, 111
Kelly, Tom, 107
Kennedy, Robert F., 113
knowledge, skills, and abilities, 51

L

Lafley, A.G., 18
Lawler, Ed, 91
leadership, 11, 18, 23, 33, 34, 51, 59,
    60, 68, 70, 74, 75, 78, 81, 85, 86,
    99, 109
    competency, 59
    development, 30, 59, 74
    organizational, 16
    skills, 60, 97
    support, 105
LEGO, 103
Legrand, Claude, 6
Lending Club, 84
Lewis and Clark, 111
lifelong learners, 31, 41, 42, 43, 49
LinkedIn, 48, 98
Lucas, George, 38

M

Madsen, Peter M., 22
management, 11, 15, 19, 22, 23, 24,
    30, 31, 34, 40, 51, 59, 73, 74,
    77, 78, 79, 81, 85, 86, 87, 93, 97,
    108, 109
    approaches, 74
    fad, 6
    idea-, 85
    processes, 6, 30, 74
    programs, 22-3, 66
    style, 84
    teams, 87
Martinuzzi, Bruna, 55
McGinnis, Alan Loy, 55
McIlvaine, Andrew, 42, 91
McKnight, William L., 15, 81
Meredith, Bill, 114
micromanagement, 27
Microsoft, 3, 53
Millennials, 27
mind mapping, 59
mistakes, 8, 13, 15, 16, 18, 24, 76,
    78, 101, 112
monetary rewards, 67
Morath, Julianne M., 17, 75
"More Stay Interviews, Fewer Exit
    Interviews," 93
motivation(s), 44, 48, 57, 61, 64, 65,
    66, 90, 107
    experts, 55
    extrinsic, 63
    intrinsic, 63-4, 67, 94
    self-, 107
Mulally, Alan, 76

N

near misses, 21, 22
Netflix, 93
Nike, 63, 98-9
nimble, 22
non-monetary awards, 68
nontangible rewards, 68
Nunes, Paul, 9

O

O'Connor, Sandra Day, 111
openness, 1, 24, 28, 37, 44, 45, 104

optimism, 47, 55, 56
optimistic, 47, 55, 56, 104
organizational
   barriers, 109
   boundaries, 63
   boxes, 90
   capability, 2
   climate, 16
   design, 27
   elements, 104, 105
   environment, 10
   functions, 108
   goals, 97
   gold, 19
   innovation, 30, 92
   learning, 17
   levels, 60
   lobotomy, 39
   members, 28, 63, 79
   needs, 59
   performance, 11
   processes, 31, 80
   resources, 10
   structure(s), 27, 82, 107
   support, 64
   survival, 7
   values, 7, 29

P

pattern recognition, 59
peer recognition, 65, 67, 68
performance appraisal, 21, 29, 87-8,
   95, 96, 97, 105, 109
performance management, 92, 97
pessimistic, 55-6
Peters, Tom, 79
Pink, Daniel H., 27, 54, 78, 79
Pisano, Gary P., 19, 21
Pixar, 38, 84
playfulness, 54, 106, 109
Pogue, Janet, 83
Porras, Jerry, 106
positive attitude, 46, 47
positive visualization, 56
Post-it Note, 14, 63, 99
Prahalad, C.K., 2, 53
PricewaterhouseCoopers (PWC), 3
problem identification skills, 58

problem-solving, 39, 41, 64, 65, 86,
   96, 110, 112, 113
Procter & Gamble (P&G), 18, 63, 76
profitability, 3
Pygmalion effect, 56

Q

Qualcomm, 98

R

realistic job preview (RJP), 48
recruiting, 11, 31, 33, 34, 36, 46,
   95, 97
recruitment, 11, 33, 34, 36, 47, 94,
   96, 97, 99
Regeneron Pharmaceuticals, 2
reinforcement theory, 62, 71
reinvention, 9
resilience, 33, 59
resilient, 19, 43
respect, 25, 70
reward and recognition program,
   61-6, 69, 70
risk(s), 9, 11, 15, 17, 27, 44, 45, 67,
   75, 80, 86, 87, 101, 104, 105,
   106, 109, 112, 113, 114
routinization, 80
ROWE, 80, 93
Ruth, Babe, 14

S

Salesforce.com, 3
Samsung, 3, 22-3
Sartain, Libby, 94
Saturn, 29
scorecard, 85
Sebra, William, 47, 90
self-awareness, 45
self-fulfilling prophecy, 56
self-starter(s), 43, 44
"Shifting Focus to Agile
   Development," 2
silo mentality, 40, 52, 91
Siuty, Luke, 36
Skype, 48
Smith, Fred, 111
spontaneity, 27
stakeholder(s), 6, 30, 49, 82, 90

Stanleigh, Michael, 3, 11, 98
strategic, 39, 40, 89
    approach, 38, 90
    business partner(s), 90
    plan, 23
    redesign, 94
    role, 91
    thinkers, 38, 39, 40, 41, 43
    thinking, 33, 38, 39, 41
    view, 53
    work, 11
Su, Tina, 54, 56

T

talent, 3, 11, 33, 34, 41, 47, 48, 50,
    60, 77, 83, 90, 96, 97
    management, 34, 47, 53, 95, 97
    managers, 92
    pipeline, 59
    retention, 99
tangible rewards, 67
Tartell, Ross, 10
team rewards, 69
teamwork, 69, 84, 99
Tinsley, Catherine H., 22
Todd, Robert, 49
Towers Watson, 90
Toyota, 74, 95
training, 11, 30, 40, 43, 49, 91, 92,
    94, 96, 99, 100, 109
transparency, 23, 24, 30
Trenam Kemker, 83
trust, 13, 15, 23, 25, 30-1, 54, 68, 76,
    81, 87, 93, 94, 104
tunnel vision, 37, 52
Twitter,

U

uncertainty, 14, 44, 70, 112, 113
Unilever Indonesia, 3
UPS, 18, 99

V

virtual teams, 26
voluntary terminations, 93

W

W.L. Gore & Associates, 19, 79
Waterman, Robert, 79
Whole Foods, 74
Whole New Mind, 54
"Why We Hate HR," 89
Wood, Robert Chapman, 26
World Bank, 64
Wright, Orville, 110
Wright, Wilbur, 110

X

Xyntéo Ltd., 26

Y

Yahoo, 94

Z

Zappos, 10, 33, 35, 97
Zuckerberg, Mark, 111

About the Author

Patricia M. Buhler, DBA, SHRM-SCP, SPHR, is full professor at Goldey-Beacom College. She teaches several undergraduate- and graduate-level management and HR courses. She is a law firm consultant on HR-related legal issues. Dr. Buhler is the author/coauthor of several books, including *Up, Down, and Sideways: High-Impact Verbal Communication for HR Professionals* (Society for Human Resource Management, 2013) with Dr. Joel D. Worden.

Additional SHRM-Published Books

The ACE Advantage: How Smart Companies Unleash Talent for Optimal Performance
William A. Schiemann

Business-Focused HR: 11 Processes to Drive Results
Scott P. Mondore, Shane S. Douthitt, and Marisa A. Carson

The Cultural Fit Factor: Creating an Employment Brand That Attracts, Retains, and Repels the Right Employees
Lizz Pellet

Defining HR Success: 9 Critical Competencies for HR Professionals
Kari R. Strobel, James N. Kurtessis, Debra J. Cohen, and Alexander Alonso

Hidden Drivers of Success: Leveraging Employee Insights for Strategic Advantage
William A. Schiemann, Jerry H. Seibert, and Brian S. Morgan

HR's Greatest Challenge: Driving the C-Suite to Improve Employee Engagement and Retention
Richard P. Finnegan

Got a Solution? HR Approaches to 5 Common and Persistent Business Problems
Dale J. Dwyer and Sheri A. Caldwell

The Power of Stay Interviews for Engagement and Retention
Richard P. Finnegan

Repurposing HR: From a Cost Center to a Business Accelerator
Carol E. M. Anderson

Transformational Diversity: Why and How Intercultural Competencies Can Help Organizations to Survive and Thrive
Fiona Citkin and Lynda Spiel